Current Topics in
217 Microbiology
and Immunology

Springer
Berlin
Heidelberg
New York
Barcelona
Budapest
Hong Kong
London
Milan
Paris
Santa Clara
Singapore
Tokyo

Molecular Analysis
of DNA Rearrangements
in the Immune System

Edited by R. Jessberger
and M.R. Lieber

With 43 Figures and 13 Tables

 Springer

Dr. Rolf Jessberger
Institute for Immunology
Grenzacherstr. 487
CH-4005 Basel
Switzerland

Dr. Michael R. Lieber
Division of Molecular Oncology
Department of Pathology
Washington University School of Medicine
Campus Box 8118, 660 So. Euclid Ave.
St. Louis, MO 63110
USA

Cover illustration: V(D)J recombination is one of three covalent alterations of DNA that generate the genes encoding antigen receptors of the immune system. Class switch recombination and somatic hypermutation are two others. All three, and related topics, are discussed in this volume.
The cover is a partial model for V(D)J recombination. The yellow circles represent the RAG-1,-2 endonuclease. This endonuclease cleaves at heptamer/nonamer signal sequences that have either a 12- or 23- base pair spacer between the heptamer and the nonamer (green and red triangles, respectively). The endonuclease generates a hairpinned coding end and a blunt signal end. Ku70/80, a heterodimer, is the predominant DNA end binding protein in cells and is thought to bind to DNA termini. In one model, Ku is the DNA binding component of a complex that includes the DNA-dependent protein kinase, DNA-PKcs. DNA-PKcs appears to be the defective component in the double-strand break repair complementation group XRCC7. XRCC7 also is the complementation group for murine SCID. Ku70 is defective in XRCC6, Ku80 is defective in XRCC5. In V(D)J recombination, the signal ends are joined to form a signal joint. The V and J sub-exons are joined to form the coding joint, creating a novel exon in the immune-specific repertoire armamentarium.

Cover design: Springer-Verlag Heidelberg, Design & Production

ISSN 0070-217X
ISBN 3-540-61037-5 Springer-Verlag Berlin Heidelberg New York

© Springer-Verlag Berlin Heidelberg 1996
Library of Congress Catalog Card Number 15-12910
Printed in Germany

The use of general descriptive names, registered names, trademarks, etc. in this publication does not imply, even in the absence of a specific statement, that such names are exempt from the relevant protective laws and regulations and therefore free for general use.

Product liability: The publishers cannot guarantee the accuracy of any information about dosage and application contained in this book. In every individual case the user must check such information by consulting the relevant literature.

SPIN: 10510675   27/3140-5 4 3 2 1 0   – Printed on acid-free paper

Preface

The vertebrate immune system is distinctive among defense systems of multicellular organisms. In addition to nonspecific immunity, it generates a randomized array of millions of antigen receptors (immunoglobulins and T-cell receptors). A subset of these receptors are critical for binding to invading microbes or biochemicals from them to tag the microbes for elimination. Three site-directed DNA modification processes are critical to this process in vertebrates. V(D)J recombination generates the array of exons that encode the antigen binding pockets. Recent work summarized in this volume describes the dissection of this process at the biochemical level. The mechanism of the reaction is now understood in considerable detail. The proteins that catalyze many steps of the process have now been identified by biochemical and genetic reconstitution and by analysis of genetic mutants defective in V(D)J recombination.

Class switch recombination is the process by which the variable domain exon of the heavy chain is changed from IgM to IgG, IgA, or IgE. Recent progress is described in the development of an extrachromosomal substrate assay system. Molecular genetic analysis of the process in transgenics is defining some of the cis sequence requirements. Biochemical assays for defining enzymatic components are also described.

In addition to exciting progress in V(D)J recombination and class switch recombination, one chapter describes recent progress in somatic hypermutation. This is the process by which affinity maturation is achieved, and it involves generation of point mutations within and downstream of the variable domain exons of the heavy and light chain genes of immunoglobulins.

All three of these systems are intriguing because of their site-directed nature. How is targeting achieved, how are the reactions carried out, what are the enzymes that catalyze them, and how are these enzymes regulated? These are among the questions examined in this monograph.

Basel   R. JESSBERGER
St. Louis, Montana   M.R. LIEBER

Contents

List of Contributors

(Their addresses can be found at the beginning of their respective chapters)

Initiation of V(D)J Recombination in a Cell-Free System by RAG1 and RAG2 Proteins

Dik C. van Gent, J. Fraser McBlane, Dale A. Ramsden,
Moshe J. Sadofsky, Joanne E. Hesse, and Martin Gellert

1 Introduction

Mature T cell receptor (TCR) and immunoglobulin (Ig) genes are assembled from separate gene segments, which are flanked by recombination signal sequences (RSS), consisting of conserved heptamer and nonamer motifs separated by a spacer region of 12 or 23 base pairs (bp). V(D)J recombination results in a precise head-to-head ligation of two signal sequences in a signal joint, and the imprecise joining of two coding segments in a coding joint, which may contain additions or deletions of a few bp. The imprecise nature of coding joints led to the hypothesis that double strand breaks (DSB) might be intermediates in the recombination pathway. These DSB can then be processed by an exonuclease and/or by terminal deoxynucleotidyl transferase (TdT) before formation of a coding joint.

Such broken molecules have indeed been found in lymphoid cells (Roth et al. 1992a). Signal ends are blunt, 5'-phosphorylated, and contain the complete signal sequence (Schlissel et al. 1993; Roth et al. 1993). Coding ends were initially found only in mice harboring the severe combined immunodeficiency (scid) mutation. They terminated in a hairpin structure in which both strands

Laboratory of Molecular Biology, National Institute of Diabetes and Digestive and Kidney Diseases, National Institutes of Health, Bethesda, MD 20892–0540, USA

were covalently coupled (ROTH et al. 1992b). These hairpin coding ends have since also been found in a non-scid pre-B cell line, albeit at a very low level (RAMSDEN and GELLERT 1995). They were hypothesized to be intermediates in normal V(D)J recombination, since this would provide a plausible explanation for the frequent observation of self-complementary (P) nucleotides: opening of the hairpin away from the tip could produce such a sequence. Additional evidence that broken molecules are indeed intermediates in V(D)J recombination was obtained from studies on the 103/BCL-2 cell line, which contains a temperature sensitive v-abl gene. Upon shift from the permissive (low) to the nonpermissive (high) temperature, the v-abl protein is inactivated, expression of RAG proteins is upregulated, and a very high level of recombination is induced (CHEN et al. 1994). Under the nonpermissive conditions, coding joints are formed readily, but DSB persist at the signal ends. When the cells are shifted back to the permissive temperature, these signal ends go on to form signal joints, showing that signal ends are really intermediates leading to signal joints (RAMSDEN and GELLERT 1995).

Expression of two novel genes, RAG1 and RAG2, was found to be necessary and sufficient to confer V(D)J recombination activity on nonlymphoid cells (SCHATZ et al. 1989; OETTINGER et al. 1990). Most evidence pointed toward a role of RAG1 and RAG2 in initiation of V(D)J recombination; mice that do not express both of these proteins do not make DSB (SCHLISSEL et al. 1993). Moreover, a mutant form of RAG1 was found with highly increased sensitivity to mutations in the heptamer and flanking coding sequence, consistent with the hypothesis that RAG1 protein interacts with this region (SADOFSKY et al. 1995).

Attempts to reconstitute complete recombination in a cell-free system have not yielded any positive results, perhaps because of the large number of proteins involved. The initial cleavage reaction was expected to require at least the presence of the two RAG proteins, but the joining of these broken molecules to form signal and coding joints requires at least several other proteins involved in DSB repair (ROTH et al. 1995). It therefore seemed more practical to reproduce only the initial step of V(D)J recombination (formation of DSB) in a cell-free system.

2 Development of a Cell-Free System

2.1 Generation of Double Strand Breaks in Pre-B Cell Nuclear Extracts

In order to develop a cell-free system for generation of DSB at RSS, we searched for a very active source of recombination factors, and an efficient technique for detection of broken DNA molecules. As the source of recombination factors we used the pre-B cell line 103/BCL-2, which had been induced to very high levels of recombination (by shifting it to the nonpermissive temperature). Nuclear extracts were prepared from these cells, and tested for cleavage activity. Broken DNA molecules were detected by the ligation-mediated polymerase chain reaction (LMPCR), in which a linker is ligated onto blunt DNA ends, followed by PCR using a locus-specific and a linker-specific primer. Since the RAG proteins had been implicated in V(D)J recombination by genetic methods, we also added these proteins (produced in recombinant expression systems) to the extracts.

We found that the extracts alone generate a very low level of DSB at the border of signal sequences, and the efficiency of this reaction can be greatly enhanced by addition of recombinant RAG1 protein. Moreover, nuclear extracts from RAG1(-/-)pre-B cells (which are not active in the cleavage assay by themselves) can be complemented by recombinant RAG1 protein, indicating that RAG1 is directly involved in the cleavage reaction. The cleavage activity from nuclear extracts was partially purified, and shown to co-fractionate with RAG2 protein, suggesting that the RAG2 protein is also directly involved in generating DSB. The same products were generated as observed previously in vivo: full length, blunt, 5'-phosphorylated signal ends, and coding ends with a hairpin structure, suggesting that hairpins are the direct result of the cleavage reaction (VAN GENT et al. 1995). In recombination substrates with two RSS (one with a 12 bp spacer and one with a 23 bp spacer) only one signal was cut, and substrates with one signal sequence were cleaved as efficiently as substrates with two, indicating that a single signal is recognized in the cleavage reaction.

2.2 Cleavage by RAG1 and RAG2 Proteins

Although these results showed that RAG1 and probably also RAG2 were involved in the cleavage reaction, we could not exclude the possibility that other factors from the extract were involved as well. We therefore investigated whether cleavage could be accomplished by RAG1 and RAG2 alone. In collaboration with the laboratory of M. Oettinger, we found that the pre-B cell nuclear extract could be replaced by recombinant RAG2 protein, showing that RAG1 and RAG2 are both necessary and sufficient for DSB formation (McBLANE et al. 1995). (This replacement had previously failed because not all preparations of recombinant RAG2 protein are active.) The frequency of DSB observed in reactions with purified RAG proteins was much higher than observed previously in nuclear extracts: up to 10% of input recombination substrate could be cleaved, as estimated from Southern blot analysis. Cleavage took place at a single signal only, suggesting that oligonucleotides with only one signal sequence might be good substrates. Indeed, a double-stranded 50-mer oligonucleotide substrate containing either a 12-signal or a 23-signal sequence was cut efficiently (Fig. 1). Cleavage again resulted in blunt, 5'-phosphorylated signal ends and hairpins at the coding ends.

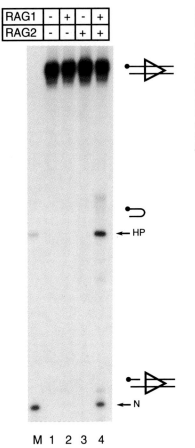

Fig. 1. Cleavage of a recombination signal sequence by purified RAG1 and RAG2 proteins. A 50 bp oligonucleotide substrate containing a signal sequence with a 12 bp spacer was incubated without proteins (*lane 1*), with RAG1 only (*lane 2*), RAG2 only (*lane 3*), or RAG1 + RAG2 (*lane 4*). A 32P label was included at the 5′ end on the coding side; products were separated on a 12.5% TBE/urea gel, and visualized by autoradiography. (*M*, marker lane, containing a 16 nucleotide marker – the position of the nicked intermediate – and a hairpin marker; *N*, DNA species nicked at the border of the signal sequence and the flanking coding DNA; *HP*, hairpin formed on the coding end)

In addition to these DSB we found a DNA species containing a nick at the 5′ end of the signal sequence (Fig. 1). Moreover, an oligonucleotide substrate with a pre-existing nick at this position is a good substrate for hairpin formation at the coding end, showing that this nicked species is an intermediate on the pathway to DSB formation. Both nicking and hairpin formation require RAG1 as well as RAG2, and an RSS.

3 Outline of the V(D)J Cleavage Reaction

The data presented above lead to the following model for initiation of V(D)J recombination (Fig. 2). After RAG1 and RAG2 recognize a signal sequence, they hydrolyze the phosphodiester bond at the 5′ end of the heptamer, resulting in a 3′-OH on the coding side, and a 5′-phosphate on the signal side. This

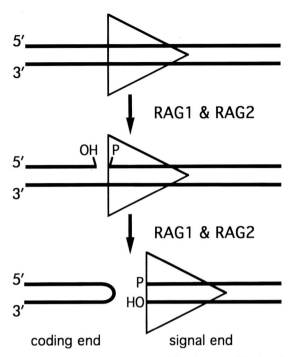

Fig. 2. Model for V(D)J cleavage. After the recombination signal sequence is recognized by the RAG1 and/or RAG2 proteins, a nick is introduced at the 5' end of the signal sequence heptamer (producing a 3' hydroxyl and a 5' phosphoryl group). The 3' hydroxyl is then coupled to the phosphate in the opposite strand, forming a hairpin structure on the coding end side. Both these reactions require RAG1 as well as RAG2. The recombination signal sequence is represented by a triangle

3'-OH is then coupled to the phosphate in the opposite strand, creating a hairpin structure at the coding end, and a blunt signal end.

Conversion of a nicked DNA species to a hairpin coding end and a blunt signal end involves breakage of one phosphodiester bond, and formation of a new one (at the tip of the hairpin). Formation of a phosphodiester bond requires energy, but no ATP or other high-energy cofactor was added to the cleavage reactions. Thus the energy of the broken phosphodiester bond is most probably conserved for the formation of the new one. The energy could be conserved by either a covalent protein-DNA intermediate (as in topoisomerases), or by direct trans-esterification. In the first case, the phosphodiester bond opposite the nick would be broken by the RAG proteins, resulting in covalent attachment of RAG1 or RAG2 to the 5' phosphoryl group, probably by coupling of a serine, threonine or tyrosine residue. Subsequently, the 3' hydroxyl produced in the initial nicking step could be used as the nucleophile to break this phospho-protein linkage, thereby releasing the protein and producing a hairpin structure on the DNA. A similar reaction has been found with several DNA recombination systems, such as bacteriophage lambda integration and the reactions catalyzed by bacterial invertases and resolvases. In the direct trans-esterification model, the 3' hydroxyl produced in the initial nicking reaction would be used directly to attack the phosphodiester bond in the opposite strand, simultaneously breaking this bond and making the new one to form a hairpin. Bacteriophage Mu transposition and retroviral integration have been found to proceed through such a mechanism: the 3'-OH produced in the initial nicking step is used as the nucleophile to break a phosphodiester bond in the acceptor DNA. Recent stereochemical experiments have shown that cleavage by the RAG proteins falls into this second class, indicating that the RAG system strongly resembles transposable elements in its chemical mechanism (VAN GENT et al. 1996).

4 Recognition of Recombination Signal Sequences

Recombination substrates with a signal sequence are cut by purified RAG1 and RAG2 proteins, showing that one or both of these proteins must be able to recognize such sequences. Several lines of evidence indicate that both RAG proteins bind to signal sequences. As mentioned above, we recently found a mutant form of RAG1 which is extremely sensitive to mutations in and around the heptamer, suggesting that RAG1 interacts with the DNA in that region. We have also found that both RAG1 and RAG2 protein have affinity for DNA, with a moderate specificity for signal sequences (M: SADOFSKY et al. unpublished results).

These results raise the question of how the two RAG proteins co-operate in recognition and cleavage of signal sequences. The simple model, in which one binds specifically and the other one is the catalytic subunit, does not seem to be very likely, since both have affinity for signal sequences. It seems more likely that both proteins contribute to the binding specificity. Although we do not have any evidence for co-operative binding, some data indicate that RAG1 and RAG2 proteins may form a specific complex. For example, RAG2 protein expressed in insect cells has a very low activity unless RAG1 is expressed in the same cells, suggesting that RAG1 can stabilize RAG2 in this environment (McBLANE et al. 1995). Furthermore, RAG2 has been found to co-immunoprecipitate with RAG1 from cells expressing both proteins (LEU and SCHATZ 1995).

RAG proteins recognize signal sequences with a 12 bp spacer or a 23 bp spacer between the conserved heptamer and nonamer sequences. The fact that both types of signal sequences are recognized by the same proteins is in agreement with our earlier findings that mutations introduced into either type of signal sequence have similar effects on recombination in transfection assays (HESSE et al. 1989). The two types of signal sequences could be recognized by different numbers of RAG protein molecules or a different conformation of the cleavage complex. It should be noted that the difference in the preferred spacer lengths is 11 bp, amounting to approximately one turn of the double helix. In other words, in both types of signals the nonamer would be located on the same face of the helix with respect to the heptamer.

Although RAG1 and RAG2 bind with some specificity to signal sequences in vitro, this binding does not seem to be specific and tight enough to explain how the RAG proteins would be able to find them in a whole genome. Specificity and affinity might be improved by formation of a specific synaptic complex (which has probably not been reproduced yet in vitro), or other factors might be involved in recruiting the RAG proteins to Ig and TCR loci.

5 Regulation of Cleavage

5.1 Coupled Cleavage at Two Signal Sequences

The cleavage reaction described here faithfully reproduces cleavage observed in vivo; the same signal and coding ends are produced. However, cleavage and recombination in vivo must be regulated at several levels which are missing in this reaction with purified proteins.

V(D)J recombination in cells requires the presence of two types of signal sequences: a signal with a 12 bp spacer always recombines with one with a 23 bp spacer (TONEGAWA 1983). The requirement for two different types of signals suggests that both may need to be present in the same synaptic complex before recombination starts. In accordance with that view, it was found that all DNA molecules broken at the TCRδ locus of new-born mice were cut at both signal sequences (ROTH et al. 1992a), and recent results with chromosomal test substrates indicate that cleavage is greatly depressed unless both a 12 and a 23 signal are present (D. B. ROTH, personal communication). However, in vitro a single signal sequence is cleaved, and even when two signals are present in one substrate, cleavages are detected predominantly at one signal sequence only.

A possible explanation for the lack of coordinated cleavage is that one or more factors may be missing from the cell-free reaction. Such a hypothetical factor could inhibit cleavage at a signal sequence by RAG1 and RAG2, unless the other signal is also present in the same cleavage complex. Alternatively, cleavage by RAG1 and RAG2 might be increased dramatically when a specific synaptic complex is formed. Such a complex could contain additional factors, required for regulation of its formation or involved in later steps of V(D)J recombination. Factors that might be present in such a complex include Ku (which binds to DNA ends and was found to be important for V(D)J recombination), the catalytic subunit of DNA dependent protein kinase (which has been identified as the scid factor), and several other proteins involved in DSB repair. Although we have not yet been able to form a synaptic complex in vitro in which cleavage is coordinated, we cannot exclude the possibility that RAG1 and RAG2 may be sufficient to form such a complex. More favorable reaction conditions or specific post-translational modifications of the RAG proteins might be required for such a reaction.

5.2 Regulation of Cleavage Activity

Because the basic cleavage reaction requires nothing beyond RAG1 and RAG2, cells must be able to avoid having them make potentially harmful DSB throughout the genome. It is to be expected that there will be extensive control of the RAG genes and proteins in vivo. One important layer of control is of course at the level of RAG gene transcription: they are transcribed in only a small subset of all cells, and their expression is tightly regulated in development (the combination of these two genes is only expressed in pre-B and pre-T cells). The fact that both RAG1 and RAG2 are required for cleavage greatly reduces the chance of accidental activity in cells that express only one of the two RAG genes but are not supposed to rearrange their Ig or TCR loci, such as in the brain (CHUN et al. 1991).

A second layer of control is at the level of stability and post-translational modification of these proteins. Both RAG1 and RAG2 have short half lives (SADOFSKY et al. 1993; LIN and DESIDERIO 1995), indicating that regulation at the level of transcription is quickly reflected at the protein level, and that cleavage activity in these cells will therefore also follow. There is evidence that the level and stability of RAG2 protein is regulated in the cell cycle, with maximal amounts being present in G1 phase (LIN and DESIDERIO 1995). This parallels formation of DSB, which are also most abundant in G1 phase (SCHLISSEL et al. 1993). Different phosphorylation forms of the RAG2 protein have been reported to have different stability or activity. Availability of the cell-free system should facilitate studies of these modified proteins.

Even in cells that are recombinationally active, only a small fraction of the RSS (or RSS-like) sequences is available at any time. This regulation, at the level of the target DNA accessibility, may also be approachable in the biochemically defined system. Access to Ig and TCR loci may be limited by their incorporation into chromatin, or by other bound protein factors. In addition to the suggested roles of transcription or DNA methylation in regulation of V(D)J recombination, some of the proteins with affinity for heptamers, or nonamers, or both (reviewed in LEWIS 1994) might also be involved in control of cleavage, acting either as activators or inhibitors. Initiation of V(D)J recombination may be further modulated by factors that bind to the RAG proteins themselves. Proteins that bind to RAG1 have been described (CUOMO et al. 1994; CORTES et al. 1994), although little evidence for a regulatory function has yet been found. Attempts to reconstitute some of these processes in vitro should now be possible.

All considerations about regulation of cleavage activity mentioned thus far depend on the assumption that RAG1 and RAG2 by themselves can make DSB in vivo. Although cleavage by these proteins appears to be reasonably efficient in vitro, this activity might not be sufficient to produce DSB in vivo without stimulatory factors. It should be noted here that efficient cleavage in the cell-free reaction requires Mn^{2+}, whereas Mg^{2+} is the predominant divalent cation in cells. With the purified proteins, cleavage in Mg^{2+} produces mainly nicked DNA, suggesting that DSB formation may require additional factor(s) with this cation.

6 A Possible Role for RAG Proteins in the Completion of the Recombination Reaction

Completion of the V(D)J recombination reaction requires opening of the coding end hairpins, addition or deletion of a few nucleotides, and formation of signal and coding joints. TdT has been shown to be required for addition of non-templated (N) nucleotides, and several DSB repair proteins are necessary for formation of signal and coding joints. It is not yet clear whether the RAG proteins only catalyze cleavage at signal sequences, or whether they might also be involved in these later stages of the reaction. One could imagine that they may be involved in opening of the hairpins or recruitment of TdT and DSB repair factors.

Acknowledgments. We thank Marjorie Oettinger and colleagues for the continuing exchange of information and ideas, and our colleagues at the Laboratory of Molecular Biology for valuable discussions. D.C.v.G. was supported by the European Molecular Biology Organization, and D.A.R. by the Medical Research Council of Canada.

Note Added in Proof. More recent work has shown that the RAG1 and RAG2 proteins produce coupled cleavage at a pair of signal sequences if the cation is changed from Mn^{2+} to Mg^{2+}. The signal pair with 12-bp and 23-bp spacers is most favored, and single-signal cleavage is strongly suppressed. Thus the RAG1 and RAG2 proteins alone are responsible for establishing the 12/23 rule in V(D)J recombination (VAN GENT DC, RAMSDEN DA, GELLERT M (1996) Cell 85:107–113).

References

Chen YY, Wang LC, Huang MS, Rosenberg N (1994) An active v-abl protein tyrosine kinase blocks immunoglobulin light-chain gene rearrangement. Genes Dev 8:688–97
Chun JJ, Schatz DG, Oettinger MA, Jaenisch R, Baltimore D (1991) The recombination activating gene-1 (RAG-1) transcript is present in the murine central nervous system. Cell 64:189–200

Cortes P, Ye Z-S, Baltimore D (1994) RAG-1 interacts with the repeated amino acid motif of the human homologue of the yeast protein SRP1. Proc Natl Acad Sci USA 91:7633–7637

Cuomo CA, Kirch SA, Gyuris J, Brent R, Oettinger MA (1994) Rch1, a protein that specifically interacts with the rag-1 recombination-activating protein. Proc Natl Acad Sci USA 91:6156–6160

Hesse JE, Lieber MR, Mizuuchi K, Gellert M (1989) V(D)J recombination: a functional definition of the joining signals. Genes Dev 3:1053–1061

Leu TMJ, Schatz DG (1995) rag-1 and rag-2 are components of a high-molecular-weight complex and association of rag-2 with this complex is rag-1 dependent. Mol Cell Biol 15:5657–5670

Lewis SM (1994) The mechanism of V(D)J joining: Lessons from molecular, immunological, and comparative analyses. Adv Immunol 56:27–150

Lin W-C, Desiderio S (1995) V(D)J recombination and the cell cycle. Immunol Today 16:279–289

McBlane JF, van Gent DC, Ramsden DA, Romeo C, Cuomo CA, Gellert M, Oettinger MA (1995) Cleavage at a V(D)J recombination signal requires only RAG1 and RAG2 proteins and occurs in two steps. Cell 83:387–395

Oettinger MA, Schatz DG, Gorka C, Baltimore D (1990) RAG-1 and RAG-2, adjacent genes that synergistically activate V(D)J recombination. Science 248:1517–1523

Ramsden DA, Gellert M (1995) Formation and resolution of double strand break intermediates in V(D)J rearrangement. Genes Dev 9:2409–2420

Roth DB, Nakajima PB, Menetski JP, Bosma MJ, Gellert M (1992a) V(D)J recombination in mouse thymocytes: double-strand breaks near T cell receptor δ rearrangement signals. Cell 69:41–53

Roth DB, Menetski JP, Nakajima PB, Bosma MJ, Gellert M (1992b) V(D)J recombination: broken DNA molecules with covalently sealed (hairpin) coding ends in scid mouse thymocytes. Cell 70:983–991

Roth DB, Zhu C, Gellert M (1993) Characterization of broken DNA molecules associated with V(D)J recombination. Proc Natl Acad Sci USA 90:10788–10792

Roth DB, Lindahl T, Gellert M (1995) How to make ends meet. Curr Biol 5:496–499

Sadofsky MJ, Hesse JE, McBlane JF, Gellert M (1993) Expression and V(D)J recombination activity of mutated RAG-1 proteins. Nucl Acids Res 21:5644–5650

Sadofsky M, Hesse JE, van Gent DC, Gellert M (1995) RAG-1 mutations that affect the target specificity of V(D)J recombination: a possible direct role of RAG-1 in site recognition. Genes Dev 9:2193–2199

Schatz DG, Oettinger MA, Baltimore D (1989) The V(D)J recombination activating gene, RAG-1. Cell 59:1035–1048

Schlissel M, Constantinescu A, Morrow T, Baxter M, Peng A (1993) Double-strand signal sequence breaks in V(D)J recombination are blunt, 5'-phosphorylated, RAG-dependent, and cell-cycle-regulated. Genes Dev 7:2520–2532

Tonegawa S (1983) Somatic generation of antibody diversity. Nature 302:575–581

van Gent DC, McBlane JF, Ramsden DA, Sadofsky MJ, Hesse JE, Gellert M (1995) Initiation of V(D)J recombination in a cell-free system. Cell 81:925–934

van Gent DC, Mizuuchi K, Gellert M (1996) Similarities Between Initiation of V(D)J Recombination and Retroviral Integration. Science 271:1592–1594

rag-1 and rag-2:
Biochemistry and Protein Interactions

DAVID G. SCHATZ[1] and THOMAS M.J. LEU[2]

1 Introduction

The process of V(D)J recombination is the defining characteristic of lymphocyte development. This site-specific recombination reaction assembles the genes that encode immunoglobulin (Ig) and T cell receptor proteins, and in many species is the source of much of the diversity in these gene products. The reaction is complex and almost certainly requires the coordinated activity of a large number of proteins. Most components of the V(D)J recombination enzymatic machinery (hereafter referred to as the V(D)J recombinase) are ubiquitously expressed, with only three lymphocyte-specific factors thus far identified: the products of the recombination activating genes (RAG-1 and RAG-2), and terminal deoxynucleotidyl transferase (TdT). TdT is not required for V(D)J recombination, but when present it functions to add nongermline-

[1] Howard Hughes Medical Institute, Section of Immunobiology, Yale University School of Medicine, 310 Cedar Street, P. O. Box 208011, New Haven, CT 06520–8011, USA
[2] Universität Zürich Irchel, Institut für Veterinärbiochemie, Winterthurerstrasse 190, 8057 Zürich, Switzerland

encoded nucleotides (N regions) to coding junctions and thereby greatly increases the diversity of the products of the reaction. The rag-1 and rag-2 proteins are central to the process of V(D)J recombination, and their biochemical properties and protein-protein interactions are the focus of this chapter.

RAG-1 and RAG-2 were isolated by combining a sensitive assay for V(D)J recombination with serial transfections of genomic DNA: in essence, searching for genes able to activate the V(D)J recombinase when transferred into non-lymphoid cells (SCHATZ and BALTIMORE 1988). RAG-1 was identified first, and presented the curious puzzle that RAG-1 cDNA clones functioned very poorly in the complementation assay that had allowed the isolation of the gene (SCHATZ et al. 1989). The puzzle was solved by the discovery of RAG-2, which was found to reside only a few kilobases away from RAG-1 (OETTINGER et al. 1990). In retrospect, it is clear that neither gene could have been isolated by the approach used had they not resided so close to one another in the genome. While RAG-1 and RAG-2 have no sequence similarity to one another, they share the same general genomic organization: the coding region is contained in a single large exon, upstream of which are one (for RAG-1; SILVER 1993) or two (for RAG-2; C. M. THOMPSON and D. G. SCHATZ, unpublished data) small exons containing portions of the 5' untranslated region. The close juxtaposition and relative orientation (convergent transcription) of the two RAG genes has been conserved in all vertebrate species from which they have been isolated and characterized (CARLSON et al. 1991; FUSCHIOTTI et al. 1993; GREENHALGH et al. 1993; ICHIHARA et al. 1992; OETTINGER et al. 1990). RAG-1 (and presumably also RAG-2) is present in all vertebrate species thus far examined (BERNSTEIN et al. 1994), but the evolutionary origin of the two genes remains unknown. It is tempting to think, based on the close proximity and simple structure of the RAG genes, that the rag proteins once functioned in a fungal or viral recombination reaction and coevolved to play their present roles in V(D)J recombination.

In keeping with the close proximity of their genes, the rag proteins cooperate closely during V(D)J recombination. Both rag-1 and rag-2 are absolutely essential for V(D)J recombination, a point first suggested by transfection studies (OETTINGER et al. 1990) and then demonstrated conclusively by targeted disruption of either RAG-1 or RAG-2 in mice (MOMBAERTS et al. 1992; SHINKAI et al. 1992). In these RAG knockout mice, lymphocyte development is arrested and those immature lymphocytes that do exist show no evidence for broken-ended molecules resulting from cleavage adjacent to recombination signal sequences (RSS), suggesting that V(D)J recombination is disrupted at a very early stage (SCHLISSEL et al. 1993; YOUNG et al. 1994). The reason for this has just been elucidated: rag-1 and rag-2 are necessary and sufficient for RSS-dependent cleavage of DNA in vitro (see chapter by D.C.VAN GENT et al., this volume). Consistent with this, we provide evidence here that rag-1 and rag-2 are components of the same large, heterogeneously-sized, protein complexes.

Little else is known about the protein interactions of rag-1 and rag-2. Use of the yeast two-hybrid system has identified two related proteins that interact with rag-1, Rch-1 (CUOMO et al. 1994) and SRP-1 (CORTES et al. 1994). These proteins likely play a role in nuclear transport of rag-1 (GORLICH et al. 1994), and their role, if any, in V(D)J recombination is unclear. It is curious that the yeast two-hybrid system failed to detect any proteins interacting with rag-2, including rag-1 (CUOMO et al. 1994). While rag-1 and rag-2 are sufficient for the early steps of V(D)J recombination (site-specific DNA binding and cleavage), it is clear that the later steps of the reaction require the direct or indirect participation of a series of DNA repair factors (see chapters by Z. LI and F. ALT, G. CHU, S.E. LEE et al., P.A. JEGGO et al., and C.W. ANDERSON and T.H. CARTER, this volume), including the components of the DNA dependent protein kinase (Ku70, Ku80 and DNA-PK$_{cs}$) and the XRCC4 gene product. It is as yet unclear whether any of these factors interact with rag-1 or rag-2, nor is it known if the rag proteins play a role in V(D)J recombination after the completion of DNA cleavage.

We have expressed the murine rag proteins at high levels in a lymphocyte cell line and compared their biochemical properties to that of rag proteins from the thymus, their richest natural source. We find that rag-1 is tightly bound in the nucleus, insoluble at physiological pH and salt concentrations, and a component of heterogeneously-sized complexes. rag-2 adopts many of these same properties when expressed in the presence of rag-1. We have developed polyclonal antibodies specific for rag-1 or rag-2, and used them to demonstrate that the rag proteins coimmunoprecipitate, thus confirming that they are components of the same complex. We also provide evidence that rag-1 (in the absence of rag-2) interacts very tightly with DNA, and that rag-1 is capable of dimerizing. The implications of the data for rag protein function and the mechanism of V(D)J recombination are discussed.

2 Expression and Biochemical Characterization of the rag Proteins

2.1 An Inducible, High Level Mammalian Expression System for rag-1 and rag-2

We chose to express the murine rag-1 and rag-2 proteins in the murine B lymphoma cell line M12 for several reasons: first, a mammalian lymphocyte cell line should provide a close physiological approximation of the normal environment of the rag proteins; second, the ability of the expressed rag proteins to perform V(D)J recombination could be determined easily by the introduction of recombination substrates into the cells; and third, the M12 cell line has been used by others to express the rag proteins using a similar expression strategy (OLTZ et al. 1993). The expression vector (LEU and SCHATZ 1995) contained a *Drosophila* heat shock protein 70 promoter to drive inducible rag expression, and a linked cassette for expression of the dihydrofolate reductase (DHFR) gene, to allow for methotrexate (MTX) selection of cells containing the vector. The full-length murine rag proteins, with a 15 amino acid N-terminal addition containing a stretch of six histidine residues, were expressed. The rag-1 and rag-2 expression vectors were transfected into M12 and the cells were subjected to several rounds of selection in increasing concentrations of MTX (intended to select for those cells with the highest DHFR expression levels, and therefore presumably the highest vector copy number and rag expression levels). Three clonal cell lines with high rag expression levels were analyzed extensively: Sr1, expressing rag-1 only; Sr2, expressing rag-2 only; and Dr3, expressing both rag-1 and rag-2 (Leu and Schatz 1995).

The levels of expression of the rag proteins in these cell lines, both before and after heat shock induction, were determined by Western blotting with rag-specific polyclonal antibodies (Leu and Schatz 1995), and by Coomassie blue staining in the case of rag-1. In addition, the absolute levels of rag-1 and rag-2 expression were estimated by immunoprecipitation followed by silver staining (Leu and Schatz 1995). Comparison of the rag protein expression levels with that observed in thymus (Table 1) reveals several significant points: first, expression levels in the transfectants before heat shock induction are higher than in thymus (even when taking into account the five- to tenfold larger volume of an M12 cell compared with a thymocyte); second, heat shock induction resulted in a 50-fold increase in rag-1 and a fivefold increase in rag-2 protein levels; and third, while the rag-1:rag-2 ratio is significantly less than 1 in the thymus, it is significantly greater than 1 in the heat shock induced Dr3 double expresser. We estimate that an average thymocyte contains approximately 10^4 molecules of rag-1, and about five times this number of rag-2 molecules. This in turn means that the transfected cell lines are expressing very substantial quantities of the rag proteins, making them useful for biochemical and enzymatic analyses of the rag proteins.

Table 1. Relative levels of the rag proteins in transfected cells and thymus

	Sr1n[a]	Sr1i	Sr2n	Sr2i	Dr3n	Dr3i	Thymus
rag-1	50	2500	NA	NA	50	2500	1[b]
rag-2	NA	NA	100	500	100	500	5
rag-1:rag-2	NA	NA	NA	NA	0.5	5	0.2

NA, not applicable.

[a] Sr1n, Sr2n and Dr3n refer to the uninduced cell lines, and Sr1i, Sr2i and Dr3i refer to the heat shock induced cell lines.

[b] All values in the table are stated relative to the number of rag-1 molecules estimated to be present, on average, in a thymocyte. This value (10^4 molecules per cell) is arbitrarily set to 1. Therefore, for example, a thymocyte is estimated to contain, on average, 5×10^4 molecules of rag-2, while a Dr3i cell is estimated to contain 5×10^6 molecules of rag-2 and 2.5×10^7 molecules of rag-1. It is worth noting that the stably transfected M12 expresser cells are bigger than thymocytes by a (volume) factor of 5–10. Data derived from Leu and Schatz 1995.

The Dr3 cell line was able to perform V(D)J recombination rapidly and at high efficiency, consistent with its high rag expression levels. Transient transfection of V(D)J recombination substrates into Dr3 resulted in a recombination frequency of between 10% and 20% (LEU and SCHATZ 1995), with the first recombination products detected 7 h after transfection using a sensitive PCR assay (D.G. SCHATZ and T.M.J. LEU, unpublished data). Therefore, the rag proteins expressed from these expression vectors are biologically active. Interestingly, heat shock induction of Dr3 did not substantially increase the recombination frequency (less than twofold), despite the substantial increase in rag protein levels (Table 1). Several, not mutually exclusive, explanations for this result seem possible: first, the rag proteins induced by heat shock may have a lower specific activity than those expressed before induction; second, the altered ratio of rag-1:rag-2 may result in the formation of nonproductive complexes; and third, the rate of recombination may become limited by the levels of some other factor(s) at very high rag expression levels.

2.2 The Subcellular Localization of the rag Proteins

As would be expected for proteins involved in recombination, immunofluorescent staining of cells with anti-rag antibodies revealed that rag-1 and rag-2 are predominantly located in the nucleus in both thymocytes and the transfected cell lines (LEU and SCHATZ 1995). Both punctate and diffuse staining were seen for both rag-1 and rag-2. In thymus, in the single rag-1 expresser Sr1, and particularly in the double expresser Dr3, anti-rag-1 antibodies yielded relatively intense staining of a small number of discrete regions of the nucleus in some cells. These regions stained poorly with propidium iodide, and subsequent double immunofluorescent staining with anti-rag-1 and anti-fibrillarin antibodies identified these regions as nucleoli (M. J. DIFILIPPANTONIO and D. G. SCHATZ, unpublished data). Interestingly, rag-2 is relatively excluded from nucleoli in the absence of rag-1 (in Sr2 cells), but is easily detected in nucleoli when coexpressed with large quantities of rag-1 in induced Dr3 cells. Therefore, rag-2 does not have any intrinsic affinity for the nucleolus, but can be localized there in a rag-1 dependent manner. The functional significance of nucleolar localization of a portion of the rag proteins is unclear, but these results provide evidence in support of an interaction in vivo between rag-1 and rag-2.

2.3 Extraction of the rag Proteins from the Nucleus

Are the rag proteins freely diffusible in and easily extracted from the nucleus, or are they tightly bound to nuclear structures? Simple nuclear extraction procedures demonstrated that the latter is the case for both rag proteins, but again, the presence of rag-1 affects the behavior of rag-2 (LEU and SCHATZ 1995). Very little rag-1 or rag-2 was found in a low salt cytoplasmic extract,

Table 2. Extraction of rag proteins from transfected cells[a]

	150 mM NaCl[b] (%)	2 M NaCl (%)	2 M + sct (%)	Residual pellet (%)
rag-1 from:				
Sr1 or Dr3[c]	5[d]	45	15	35
rag-2 from:				
Sr2	10	80	10	0
Dr3	5	60	20	15

[a]The numbers represent the percentage of the indicated rag protein obtained in the supernatant after the indicated extraction procedure, or, for the last column, the amount of protein found in the residual pellet. Data derived from LEU and SCHATZ 1995
[b]The extractions performed were, in this order, 150 mM NaCl cytoplasmic extract, 2 M NaCl nuclear extract, and re-extraction of pelleted nuclear material in the presence of 2 M NaCl combined with ultrasonic treatment (2 M + sct). The residual pellet represents the insoluble material after the last extraction. For further details, see LEU and SCHATZ 1995
[c]rag-1 behaves similarly when extracted from Sr1 or Dr3.
[d]Yield strongly dependent on speed and duration of centrifugation.

and less than half of the rag-1 protein could be extracted with 2 M NaCl (Table 2). Subsequent sonication in the presence of 2 M NaCl failed to extract all of the remaining rag-1, leaving significant quantities in the insoluble pellet (Table 2). Resistance to extraction with high salt is a characteristic of components of the nuclear matrix, and it is likely that a portion of rag-1 is tightly bound to the matrix. rag-2 from Dr3 cells behaved much like rag-1, but the rag-2 from Sr2 cells behaved differently, with a substantial increase in the fraction of the protein extracted by 2 M NaCl, and with virtually none left in the residual pellet (Table 2). Thus the behavior of rag-2 was altered by the presence of rag-1, again arguing for an interaction between the proteins. The behavior of rag-1 (and of rag-2 in the presence of rag-1) was heterogeneous rather than discrete: one population was extracted at 2 M NaCl, another required sonication and high salt, and yet a third remained tightly bound to residual nuclear components.

2.4 The Solubility of the rag-1 Is Strongly Influenced by Salt and pH

The difficulty in extracting rag-1 from the nucleus, as well as numerous attempts to purify rag-1, strongly suggested that the protein was poorly soluble at physiological pH and salt concentrations. This was confirmed to be the case by diluting high salt extracts containing rag-1 into buffers at different pH values, at either 100 mM or 600 mM NaCl (LEU and SCHATZ 1995). At the lower salt concentration, the pH had to be raised to a value of 8 before even small quantities of rag-1 remained soluble, and only at pH 11 was the majority of rag-1 soluble (Fig. 1). In contrast, rag-1 was soluble at all pH values in 600 mM

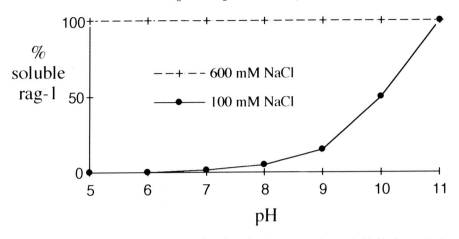

Fig. 1. Solubility of the rag-1 protein as a function of salt concentration and pH. Nuclear extract from induced rag-1 single expresser cells (Sr1i) was prepared using ultrasonication in the presence of 600 m*M* NaCl, pH 6.8. The nuclear extract was diluted into various buffers to obtain the indicated salt (600 m*M* or 100 m*M* NaCl) and pH (from 5 to 11) conditions. The samples were spun (15 min at 14 000 *g*) and the amount of rag-1 protein remaining in the supernatant was determined by Western blotting and is plotted against pH. Data derived from LEU and SCHATZ (1995)

salt. The poor solubility of the full-length rag-1 protein under physiological conditions complicates attempts to purify the protein and to perform assays with it. We and others (VAN GENT et al. 1995) have found that a smaller fragment of rag-1 (the "core" region, defined as the minimal region required for recombination activity–see Sect. 4) is more readily soluble and is considerably easier to manipulate. It is worth noting that the nuclear extraction and solubility properties of the rag proteins are the same regardless of whether the proteins are obtained from the thymus, or from uninduced or induced transfected cells. Therefore, these results are not an artifact of overexpression or heat shock induction.

2.5 The rag Proteins Are Components of Large, Heterogeneously Sized Complexes

The results described above demonstrate that the rag proteins are difficult to extract from the nucleus and (for rag-1) poorly soluble under physiological conditions. We wished to determine if the rag proteins, extracted and maintained in high salt (600 m*M* NaCl), were in complexes or behaved as protein monomers. Protein extracts from heat shock induced Sr1, Sr2, and Dr3 cells were fractionated on 15%–50% glycerol gradients in the presence of 2 *M* NaCl and the fractions assayed by Western blot for the presence of the rag proteins (LEU and SCHATZ 1995). A minority of rag-1 (expected molecular mass 120 kDa) cofractionated with the 160-kDa marker protein, with a relatively

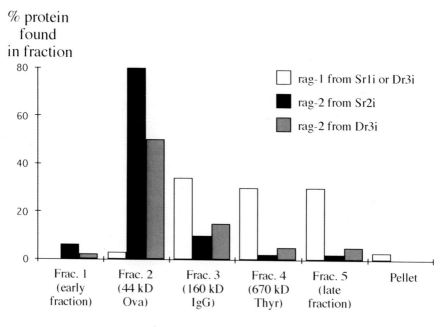

% protein
found
in fraction

glycerol gradient fraction

Fig. 2. Fractionation of the rag proteins on high salt glycerol density gradients. Nuclear extracts from induced double expresser cells (Dr3i), single rag-1 expresser cells (Sr1i), or single rag-2 expresser cells (Sr2i) were prepared in the presence of 2 *M* NaCl without ultrasonication. The extracts were centrifuged (14 h, 166 000 *g*, 4°C) through 15%–50% linear glycerol gradients in the presence of 2 *M* NaCl, and the gradients were separated into five fractions and the residual pellet at the bottom of the gradient. The amount of the rag proteins in each fraction was determined by Western blotting. Ovalbumin (44 kDa), gamma globulin (158 kDa) and thyroglobulin (670 kDa) were mixed into the extracts prior to loading on the glycerol gradients. Ovalbumin (*Ova*) was found predominantly in fraction 2, immunoglobulin (*IgG*) in fraction 3 and thyroglobulin (*Thyr*) in fraction 4. Data derived from LEU and SCHATZ (1995)

large amount of the protein found in the 670-kDa marker fraction as well as the late fraction and small amounts in the pellet at the bottom of the gradient (Fig. 2). Similar behavior was seen for rag-2 in the presence of rag-1: only about half of rag-2 (expected molecular mass 60 kDa) was found in the fraction with the 44-kDa marker and clearly detectable amounts were found in the subsequent fractions (with very little in the pellet). In the absence of rag-1, however, there was a substantial increase in the amount of rag-2 in the 44-kDa fraction (Fig. 2). Therefore, rag-1 is a component of large complexes of heterogeneous size, and rag-2 adopts this behavior in the presence of rag-1. In the absence of rag-1, much of rag-2 fractionates as expected for a monomer. This behavior was also observed for rag proteins extracted from thymus or uninduced Dr3 cells, arguing that it is not a consequence of overexpression or heat shock induction. It is also worth noting that if an extract from Dr3 cells was heated in the presence of sodium dodecyl (SDS) to denature the complexes, both rag-1 and rag-2 fractionated as expected for monomeric proteins, demonstrating that the complexes are not irreversible aggregates.

3 Associations of the rag Proteins

The data above demonstrate that rag-1 influences the behavior of rag-2 in several respects: nuclear localization, ease of extraction from the nucleus, and participation in large complexes in cellular extracts. This, together with their intimate functional association, suggested that the rag proteins might well be components of the same protein complex or complexes. We were able to test this directly using the anti-rag antibodies to immunoprecipitate the rag proteins. These experiments confirmed that rag-1 and rag-2 are components of the same complex, provided some insight into the relative stoichiometry of the association, and revealed two distinct populations of rag-1 molecules (LEU and SCHATZ 1995).

3.1 Coimmunoprecipitation of the rag Proteins

High salt detergent extracts were prepared from uninduced and induced transfected cell lines, and from thymus, and subjected to immunoprecipitation with either anti-rag-1, anti-rag-2, or control rabbit IgG antibodies (Fig. 3). Each anti-rag antibody is able to precipitate its cognate rag protein. In addition, anti-rag-1 antibodies were able to coprecipitate rag-2 from thymus or Dr3 extracts but not from Sr2 cells (Fig. 3a, compare lanes 2, 11 and 14 with lane 8), demonstrating the specificity of the antibodies. In contrast, anti-rag-2 antibodies failed to precipitate detectable quantities of rag-1 from thymus, and precipitated only small amounts from uninduced or induced Dr3 cells (Fig. 3a, lanes 3, 12, and 15). Therefore, a salt- and detergent-stable complex or complexes exist containing both rag-1 and rag-2, and there is a sharp asymmetry in the ability of the anti-rag antibodies to coimmunoprecipitate the opposite rag protein.

Several possible explanations for this asymmetry exist: first, the anti-rag-2 antibodies might disrupt the rag-1:rag-2 complex; second, the anti-rag-2 antibodies might bind to a portion(or portions) of rag-2 that is obscured in the complex; and third, the complexes might contain multiple molecules of rag-2 for each molecule of rag-1. The first possibility is unlikely because multiple, sequential immunoprecipitations with anti-rag-2 antibodies leave substantial amounts of the rag-containing complex behind in the extract (as detected by immunoprecipitation with anti-rag-1 antibodies). This result supports the second possibility by demonstrating that the anti-rag-2 antibodies immunoprecipi-

A   Immunoprecipitations

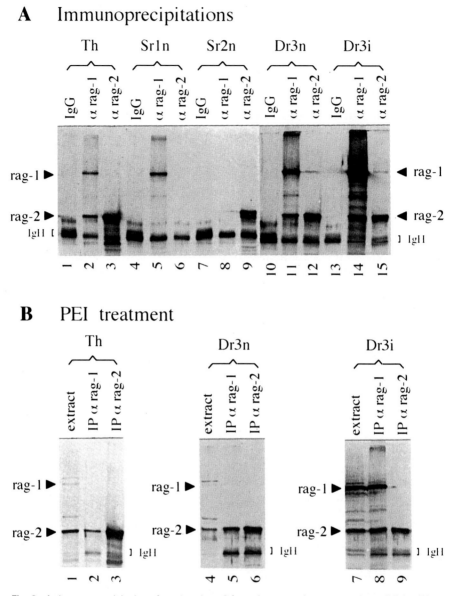

B   PEI treatment

Fig. 3 a,b. Immunoprecipitation of rag-1 and rag-2 from thymus and rag-expressing cell lines. Western blots are shown. **a** Total extracts of thymic tissue (*Th*) and cells were prepared in the presence of 1 *M* NaCl and using ultrasonication(*Sr1n*, non induced rag-1 single expresser cells; *Sr2n*, non induced rag-2 single expresser cells; *Dr3n*, non induced double expresser cells; *Dr3i*, induced double expresser cells). The extracts were precleared with rabbit immunoglobulin (IgG, coupled to agarose beads). For the immunoprecipitations (IP) antibodies were used as indicated (rabbit IgG, αx rag-1 or αx rag-2, all coupled to agarose beads). The antibody beads were washed and bound material was eluted with sample buffer (also eluting off some of the coupled antibodies, visible on the Western blots in form of heavy chain (IgH). Eluates equivalent to either 4 mg tissue (Th), 2 x 10^6 cells (Sr1n, Sr2n and Dr3n) or 2 x 10^5 cells (Dr3i) were separated in each lane of a 6.5% SDS-polyacrylamide gel (SDS-PAG).

tate free rag-2 more efficiently than rag-2 in the complex. Evidence presented in the next section supports the third possibility, namely, that the ratio of rag-1 to rag-2 in the complex is less than 1. Therefore we believe that the asymmetry of coimmunoprecipitation is explained by a combination of the second and third possibilities.

3.2 Two Distinct Populations of rag-1

As described in Sect. 2.3 and shown in Table 2, rag-1 is only efficiently extracted from nuclei following sonication, and most of our experiments were performed with extracts made with sonication. One consequence of this is that the extracts contained large amounts of DNA and RNA, and this might substantially influence the behavior of the rag proteins. To investigate this, extracts from thymus, or uninduced or induced Dr3 cells were treated with polyethylenimine (PEI), a polycation that efficiently binds to and precipitates nucleic acids. In an attempt to ensure that the PEI treatment removed only nucleic acids, it was performed in the presence of 1.2 M NaCl, a sufficiently high salt concentration to prevent direct interaction of PEI with proteins and of most proteins with DNA (BURGESS 1991). After removal of the PEI:nucleic acids precipitate by centrifugation, the supernatant was directly analyzed by Western blot (Fig. 3b, lanes 1, 4, and 7) or subjected to immunoprecipitation with the anti-rag antibodies (remainder of Fig. 3b) (LEU and SCHATZ 1995). PEI precipitation removed roughly 10% of the rag-1 protein from the thymus extract and 50%–70% from the Dr3 extracts, but had no effect on the amount of rag-2 (Fig. 3b, lanes 1, 4, and 7; and see LEU and SCHATZ 1995). Therefore, even at 1.2 M NaCl, some rag-1, but no rag-2, is associated with nucleic acids. Strikingly, anti-rag-1 antibodies precipitated as much rag-2, but substantially less rag-1, as compared with extracts that had not been treated with PEI (Fig. 3b, lanes 2, 5, and 8; compare with Fig. 3a, lanes 2, 11, and 14). The amount of immunoprecipitated rag-1 was below the level of detection for thymus and uninduced Dr3 cell extracts despite the clear signal for rag-2 (Fig. 3b, lanes 2 and 5). We emphasize that coimmunoprecipitation of rag-2 with anti-rag-1 antibodies is strictly dependent on the presence of rag-1 and

Fig. 3 b Total extracts (without IgG preclearing) were supplemented to 1.2 M NaCl and 0.5% polyethylenimine (*PEI*). The PEI-treated supernatants (extract) were used for IP, performed as described in **a**. An equivalent of either 0.04 mg tissue (Th extract), 4 mg tissue (Th IP αx rag-1 and Th IP αx rag-2), 2×10^4 cells (Dr3n extract and Dr3i extract), 2×10^6 cells (Dr3n IP αx rag-1 and Dr3n IP αx rag-2) or 2×10^5 cells (Dr3i IP αx rag-1 and Dr3i IP αx rag-2) were analyzed an a 6.5% SDS-PAG. Reprinted by permission of the American Society for Microbiology from LEU and SCHATZ 1995

is not due to antibody cross reactivity or nonspecific precipitation. These results strongly support the contention that the ratio of rag-1 to rag-2 in the complexes is less than 1. It appears therefore that two distinct populations of rag-1 exist in these extracts, and perhaps in living cells as well: the first is extremely tightly associated (salt stable but SDS labile) with nucleic acids but not with rag-2, and the second is in a complex with rag-2 but is not tightly associated (salt and SDS labile) with nucleic acids. This is discussed further below.

3.3 rag-1 Dimers?

In addition to the expected band at approximately 120 kDa, the rag-1 specific antibodies typically detected a variety of other molecular weight species in Western blots of extracts from induced expresser cells and when rag-1 was enriched by immunoprecipitation from either thymus or expresser cells (LEU and SCHATZ 1995). Two groups of bands were seen above 120 kDa: several bands above 200 kDa, referred to as rag-1m, and a ladder of 3–5 bands between 120 and 160 kDa, referred to as rag-1n. In addition, a predominant breakdown product of approximately 90 kDa, referred to as rag-1b, was often seen. These bands were only detected in extracts containing rag-1, and therefore do not represent cross-reactivity with other proteins.

Western blotting of immunoprecipitations using the anti-rag-1 antibodies and lysates from Dr3n cells, containing substantial quantities of the 90-kDa rag-1b breakdown band, revealed two new rag-1m bands of apparent molecular weight of 180 kDa and 210 kDa, as well as the band at approximately 240 kDa (Fig. 4, lane 3). Two features of these bands suggest that they represent rag-1 dimers. First, they are of the expected sizes for the possible dimers that could form between full-length rag-1 and rag-1b (90+90, 90+120, and 120+120 kDa). Second, they are in the expected relative amounts (240 kDa 210 kDa 180 kDa), given that the amount of full-length rag-1 is greater than the amount of rag-1b (Fig. 4, lane 4). Additional support for rag-1 dimerization comes from a comparison of the sizes of the rag-1m signals between thymus and Dr3 cells (Fig. 4, lanes 1 and 2). The N-terminal addition to rag-1 in the Dr3 cells gives a slight increase in the apparent molecular mass such that thymic rag-1 migrates slightly faster, and this difference is seen both in the rag-1 monomers and in the rag-1m bands. We do not know as yet whether the dimer subunits associate covalently or noncovalently; in either case, at least part of the association is able to survive boiling in SDS under reducing conditions, followed by standard SDS polyacrylamide gel electrophoresis (SDS/PAGE). Given the results of glycerol gradient fractionation (Sect. 2.5 and Fig. 2), it is possible that a majority of rag-1 protein molecules exist in the form of dimers (or higher order multimers).

Because the anti-rag-1 antibodies used in this experiment recognize rag-1b and are specific for a region near the N-terminus of rag-1 (amino acids 56–123), the degradation that gives rise to rag-1b must occur predominantly at the

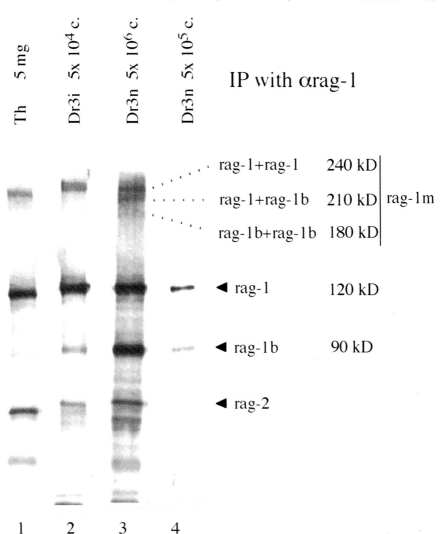

Fig. 4. Evidence for the formation of rag-1 dimers. Extracts from 5 mg thymus (*Th*) tissue, or the indicated number of uninduced (*Dr3n*) or induced (*Dr3i*) double expresser cells was subjected to immunoprecipitation with the anti-rag-1 antibodies, and the immunoprecipitated rag proteins detected by Western blotting with both anti-rag-1 and anti-rag-2 antibodies. The positions of full-length rag-1 and rag-2 proteins are indicated, as are the positions of the predominant breakdown product of rag-1 (*rag-1b*) and the higher molecular weight forms of rag-1 (*rag-1m*). Three rag-1m bands are detected in lane 3, of approximately the sizes expected for rag-1:rag-1, rag-1:rag-1b and rag-1b:rag-1b dimers. Note that rag-1, rag-1b and the rag-1m proteins from Dr3 are larger than those from thymus due to the N-terminal extension present on the rag-1 protein expressed from the transfected rag-1 expression vector

C-terminus, and the putative dimerization domain must reside in the N-terminal two-thirds of rag-1. The N-terminal one-third of rag-1 contains a domain that forms stable dimers when expressed in and purified from bacteria (K. K. RODGERS and J. E. COLEMAN, personal communication) providing additional support for the existence of rag-1 dimers.

4 Discussion

In the first few years after the isolation of their genes, relatively little was learned about the physical and enzymatic properties of the rag proteins, due to a considerable extent to difficulties in their purification. The results described here underscore the challenges of working with the full-length rag proteins: they are difficult to extract from cells, they are relatively insoluble under physiological conditions (particularly rag-1), and they are components of large complexes of heterogeneous size. We have had particular difficulty in finding conditions that allow column fractionation of rag-1: even under conditions of high salt, the protein appears to bind irreversibly to the column material.

An important step forward in our ability to manipulate the rag proteins came as a result of mutagenesis studies which defined the minimal, or "core", regions of rag-1 (SADOFSKY et al. 1994b; SILVER et al. 1993) and rag-2 (CUOMO and OETTINGER 1994; SADOFSKY et al. 1994a) required for enzymatic activity. Nearly 40% of rag-1 (amino acids 1–383 and 1009–1040) and 25% of rag-2 (amino acids 383–527) can be deleted, and the remaining proteins are able to carry out V(D)J recombination of transiently transfected substrates. The core rag proteins are substantially more soluble under physiological conditions than the full-length rag proteins (Q. EASTMAN and D. G. SCHATZ, unpublished data), and hence are more amenable to purification. The use of core rag-1 purified from insect cells contributed to the recent major breakthrough in the study of V(D)J recombination: site-specific, in vitro cleavage of DNA adjacent to RSSs (VAN GENT et al. 1995). The Gellert and Oettinger laboratories have gone on to demonstrate that a mixture of pure core rag-1 and pure core rag-2 is highly active in this cleavage assay (see chapter by D.C.VAN GENT et al., this volume). It seems clear that a great deal will be learned about the enzymatic properties of the rag proteins and the mechanism of V(D)J recombination in the near future.

4.1 Useful Reagents for the Study of the rag Proteins

We have developed murine B lymphoma cell lines that express high levels of rag-1 and/or rag-2, and in which expression can be induced by a factor of 5–100 by heat shock. Expression levels, both uninduced and induced, have remained stable even after many months of continuous culture, and the proteins expressed are active in performing V(D)J recombination. This inducible expression system has also been used to express the core rag-1 and core rag-2 proteins, with equal success. The core rag proteins from these cells perform site-specific cleavage of DNA in vitro at high efficiency (Q. EASTMAN, T.M.J. LEU and D.G. SCHATZ). While more labor intensive than some other expression methods, this system has certain advantages, including stable, high level, reproducible expression in murine lymphocytes. We have also developed polyclonal antibodies specific for rag-1 or rag-2. These antibodies are specific, high affinity reagents that are effective for detection of the rag proteins by Western blotting, immunoprecipitation and immunostaining of cells and tissues.

4.2 The Composition of rag Protein Complexes

We have shown that rag-1 affects the properties of rag-2 in several regards, altering its ease of extraction from the nucleus, its nuclear localization and its glycerol gradient fractionation properties. The existence of complexes containing rag-1 and rag-2 was confirmed by immunoprecipitation experiments, in the course of which there arose two pieces of evidence to suggest that rag-2 is more abundant in these complexes than rag-1. First, anti-rag-1 antibodies much more efficiently coimmunoprecipitate rag-2 than anti-rag-2 antibodies coimmunoprecipitate rag-1 (Fig. 3a). Second, after removal of nucleic acids (and some rag-1 protein) by precipitation with PEI, anti-rag-1 antibodies precipitate perhaps 3–5 times more rag-2 molecules than rag-1 molecules. Interestingly, thymocytes contain about this same fivefold excess of rag-2 over rag-1 (Table 1). Taken together with data supporting the existence of rag-1 dimers (Fig. 4), these results lead us to hypothesize that in vivo, the biologically active rag protein complex consists of (at minimum) two molecules of rag-1 and perhaps six or more molecules of rag-2. Our data do not allow us to determine whether the interaction between rag-1 and rag-2 is direct or indirect, but the result from the Gellert and Oettinger laboratories that rag-1 and rag-2 are sufficient for site-specific DNA cleavage argues that the interaction is direct.

Substantial evidence (summarized by LEWIS 1994) suggests that V(D)J recombination occurs in two distinct phases: (1) RSS-recognition, synapsis and cleavage; and (2) modification, repair and ligation of the broken DNA ends. The two phases may be mediated by quite distinct protein complexes: the first, by rag-1 and rag-2, and the second, by components of the double-strand break repair machinery, likely including Ku70, Ku80, DNA-PK$_{cs}$, and the product of the XRCC4 gene, as well as TdT when it is present. It is not known if rag-1 and/or rag-2 perform any function during phase 2 of the reaction. If they do not, then the "V(D)J recombinase" is likely to be two completely distinct enzymatic entities that act sequentially. If the rag proteins do participate in the second phase, then they may associate with components of the DNA repair machinery, particularly in the presence of the cleaved DNA substrate. Such protein complexes may be difficult to identify in vivo because developing lymphocytes contain only small numbers of rag-generated broken DNA ends per cell. The existence of an efficient in vitro system for substrate cleavage should facilitate the search for these interactions.

4.3 What Might the "Nonessential" Regions of rag-1 Do?

The core of rag-1 (defined here as amino acids 383–1008 of murine rag-1), is biologically active for recombination of episomal recombination substrates in vivo (SADOFSKY et al. 1994b; SILVER et al. 1993) and for site-specific cleavage in vitro (VAN GENT et al. 1995). Furthermore, it contains the regions of rag-1 necessary for interaction with rag-2 (C. McMAHAN, M. SADOFSKY and D.G. SCHATZ, unpublished data). It has not, however, been tested for its ability to recombine endogenous V, D, and J gene segments or to support normal lymphocyte development. It seems unlikely that the large portions of rag-1 deleted to produce the core have no function. While the core represents the most highly conserved region of the rag-1 protein, the N-terminus contains several well conserved areas. In support of a function for the N-terminus of rag-1, we have found that core rag-1 is quantitatively less active than a nearly full-length version of rag-1 containing amino acids 1–1008 (referred to as R1A) in the standard in vivo recombination assay. This is true despite the fact that core rag-1 is expressed at substantially higher levels than R1A (C. McMAHAN and D.G.SCHATZ, unpublished data). Therefore, addition of amino acids 1–382 increases recombination activity and decreases expression levels, resulting in a protein with a substantially higher specific activity.

The N-terminal 382 amino acid domain of rag-1 contains a putative nuclear localization signal (amino acids 141–146) and includes the region of rag-1 (amino acids 1–288) that interacts with SRP-1 (CORTES et al. 1994), a protein involved in nuclear transport. Failure to localize to the nucleus is unlikely to account fully for the lower specific activity of the rag-1 core, however, because mutation of the putative nuclear localization signal does not interfere with nuclear accumulation of rag-1 (SILVER et al. 1993), and the rag-1 core and R1A accumulate equally efficiently in the nucleus (SADOFSKY et al. 1994b). The N-terminal region also contains a ring finger domain (amino acids 286–334; LOVERING et al. 1993), the function of which has not been reported, but which has been speculated to be involved in protein-DNA interactions. Finally, the N-terminal region of rag-1 contains a presumptive dimerization domain (K.K. RODGERS and J.E. COLEMAN, personal communication) Therefore it is conceivable that the rag-1 core interacts with the DNA substrate or forms protein complexes in a manner sub optimal for catalysis. Further mutagenesis and functional studies should help resolve this issue.

4.4 DNA Binding and Populations of rag-1

Precipitation of nucleic acids with PEI precipitates a portion of rag-1 but does not reduce the amount of rag-2 that can subsequently be coimmunoprecipitated with rag-1 (Fig. 3b). The simplest explanation for this is that rag-1 molecules that do not interact with rag-2 instead bind to DNA very tightly and are therefore precipitated with the DNA by PEI. Full-length rag-1 binds nonspecifically to DNA in vitro (unpublished results), as do small portions of the protein when expressed in bacteria (K.K. RODGERS, Q. EASTMAN, T.M.J. LEU, D.G. SCHATZ and J.E. COLEMAN, unpublished data). As might be expected, much less rag-1 is precipitated by PEI in thymic extracts, where the ratio of rag-2:rag-1 is high (and presumably most of the rag-1 is bound to rag-2), than in extracts of the double-expresser cell line Dr3, where the ratio of rag-2:rag-1 is low and much of the rag-1 cannot bind to rag-2 (see Table 1). We speculate that binding of rag-2 to rag-1 not only leads to the formation of an enzymatically active complex, but also functions to sequester or protect rag-1 from tight, nonspecific binding to DNA. Interestingly, rag-2 also causes a substantial decrease in the half-life of rag-1 (U. GRAWUNDER, D.G. SCHATZ, T.M.J. LEU, A.G. ROLINK, and F. MELCHERS, unpublished data), a mechanism that may have evolved to ensure that immature lymphocytes maintain an excess of rag-2 despite having significantly more RAG-1 mRNA than RAG-2 mRNA.

As discussed in the chapter by D.C. VAN GENT et al., this volume, rag-1 and rag-2 must be responsible for sequence-specific recognition of the RSS. After performing DNA cleavage, do rag-1 and rag-2 release from the RSS? In at least some cases, the answer may be no. Broken DNA molecules terminating in RSSs (referred to as signal-end molecules) accumulate to significant levels in normal thymocytes (ROTH et al. 1992). Interestingly, signal-end molecules disappear from thymocytes in the same developmental transition as do rag-1 and rag-2 (F. LIVAK and D.G. SCHATZ, unpublished data). The data are consistent with the idea that the rag proteins form a fairly stable association with the signal ends and thereby stabilize the broken DNA molecules. It is interesting to note that developing thymocytes contain vastly more molecules of rag-1 and rag-2 (10^4–10^5 per cell) than the number of recombination events that occur in the lifetime of a cell. While there are numerous possible explanations for this, it makes conceivable the idea that individual molecules of rag-1 and rag-2 are not in fact capable of participating in multiple recombination reactions, but rather are consumed in the reaction.

Acknowledgements. We would like to thank S. Bennett and L. Corbett for excellent technical assistance, and G. Rathbun and F. Alt for providing M12 and for advice on heat shock induction. We would also like to thank K. Rodgers, J. Coleman, U. Grawunder, A. Rolink, and F. Melchers for sharing data with us before publication. Finally, we would like to thank all the members of the Schatz lab for their advice and input during this work. T. M. J. L. was supported in part by funds administered by the Union Bank of Switzerland. This work was supported by grant #AI32524 to D. G. S. from the National Institutes of Health. D. G. S. was supported by the Howard Hughes Medical Institute.

References

Bernstein RM, Schluter SF, Lake DF, Marchalonis JJ (1994) Evolutionary conservation and molecular cloning of the recombinase activating gene 1. Biochem Biophys Res Commun 205:687–692

Burgess RR (1991) Use of polyethylenimine in purification of DNA-binding proteins. Methods Enzymol 208:202–209

Carlson LM, Oettinger MA, Schatz DG, Masteller EL, Hurley EA, McCormack WT, Baltimore D, Thompson CB (1991) Selective expression of RAG-2 in chicken B cells undergoing immuno-globulin gene conversion. Cell 64:201–208

Cortes P, Ye ZS, Baltimore D (1994) RAG-1 interacts with the repeated amino acid motif of the human homologue of the yeast protein SRP1. Proc Natl Acad Sci USA 91:7633–7637

Cuomo CA, Kirch SA, Gyuris J, Brent R, Oettinger MA (1994) Rch1, a protein that specifically interacts with the RAG-1 recombination-activating protein. Proc Natl Acad Sci USA 91:6156–6160

Cuomo CA, Oettinger MA (1994) Analysis of regions of RAG-2 important for V(D)J recombination. Nucleic Acids Res 22:1810–1814

Fuschiotti P, Harindranath N, Mage RG, Mccormack WT, Dhanarajan P, Roux KH (1993) Recombination activating genes-1 and genes-2 of the rabbit – cloning and characterization of germline and expressed genes. Mol Immunol 30:1021–1032

Gorlich D, Prehn S, Laskey RA, Hartmann E (1994) Isolation of a protein that is essential for the first step of nuclear protein import. Cell 79:767–778

Greenhalgh P, Olesen C, Steiner LA (1993) Characterization and expression of recombination activating genes (rag-1 and rag-2) in Xenopus-laevis. J Immunol 151:3100–3110

Ichihara Y, Hirai M, Kurosawa Y (1992) Sequence and chromosome assignment to 11p13-p12 of the human RAG genes. Immunol Lett 33:277–284

Leu MJ, Schatz DG (1995) rag-1 and rag-2 are components of a high molecular weight complex and association of rag-2 with this complex is rag-1 dependent. Mol Cell Biol 15:5657–5670

Lewis SM (1994) The mechanism of V(D)J joining: lessons from molecular, immunological, and comparative analyses. Adv Immunol 56:27–150

Lovering R, Hanson IM, Borden KLB, Martin S, O'Reilly NJ, Evan GI, Rahman D, Pappin DJC, Trowsdale J, Freemont PS (1993) Identification and preliminary characterization of a protein motif related to the zinc finger. Proc Natl Acad Sci USA 90:2112–2116

Mombaerts P, Iacomini J, Johnson RS, Herrup K, Tonegawa S, Papaioannou VE (1992) RAG-1-deficient mice have no mature B and T lymphocytes. Cell 68:869–877

Oettinger MA, Schatz DG, Gorka C, Baltimore D (1990) RAG-1 and RAG-2, adjacent genes that synergistically activate V(D)J recombination. Science 248:1517–1523

Oltz EM, Alt FW, Lin WC, Chen JZ, Taccioli S, Desiderio S, Rathbun G (1993) A V(D)J recombinase-inducible B-cell line – role of transcriptional enhancer elements in directing V(D)J recombination. Mol Cell Biol 13:6223–6230

Roth DB, Nakajima PB, Menetski JP, Bosma MJ, Gellert M (1992) V(D)J recombination in mouse thymocytes–double-stranded breaks near T-cell receptor delta rearrangement signals. Cell 69:41–53

Sadofsky MJ, Hesse JE, Gellert M (1994) Definition of a core region of RAG-2 that is functional in V(D)J recombination. Nucleic Acids Res 22:1805–1809

Sadofsky MJ, Hesse JE, McBlane JF, Gellert M (1994) Expression and V(D)J recombination activity of mutated RAG-1 proteins. Nucleic Acids Res 22:550–550

Schatz DG, Baltimore D (1988) Stable expression of immunoglobulin gene V(D)J recombinase activity by gene transfer into 3T3 fibroblasts. Cell 53:107–115

Schatz DG, Oettinger MA, Baltimore D (1989) The V(D)J recombination activating gene (RAG-1). Cell 59:1035–1048

Schlissel M, Constantinescu A, Morrow T, Baxter M, Peng A (1993) Double-strand signal sequence breaks in V(D)J recombination are blunt, 5'-phosphorylated, RAG-dependent, and cell cycle regulated. Genes Dev 7:2520–2532

Shinkai Y, Rathbun G, Kong-Peng L, Oltz EM, Stewart V, Mendelsohn M, Charron J, Datta M, Young F, Stall AM, Alt FW (1992) RAG-2-deficient mice lack mature lymphocytes owing to inability to initiate V(D)J rearrangement. Cell 68:855–867

Silver DP (1993) Studies of the structure and function of the RAG-1 and RAG-2 genes. PhD Thesis, Massachusetts Institute of Technology

Silver DP, Spanopoulou E, Mulligan RC, Baltimore D (1993) Dispensable sequence motifs in the RAG-1 and RAG-2 genes for plasmid-V(D)J recombination. Proc Natl Acad Sci USA 90:6100–6104

van Gent DC, McBlane JF, Ramsden DA, Sadofsky MJ, Hesse JE, Gellert M (1995) Initiation of V(D)J recombination in a cell-free system. Cell 81:925–934

Young F, Ardman B, Shinkai Y, Lansford R, Blackwell TK, Mendelsohn M, Rolink A, Melchers F, Alt FW (1994) Influence of immunoglobulin heavy- and light-chain expression on B-cell differentiation. Genes Dev 8:1043–1057

Regulation of Recombination Activating Gene Expression During Lymphocyte Development

Ulf Grawunder[1,2], Thomas H. Winkler[1] and Fritz Melchers[1]

1 Introduction

The variable domains of immunoglobulin (Ig) and T cell receptor (TCR) protein chains are encoded by a variety of V (variable), D (diversity – only occurring within IgH and TCR β and δ gene loci) and J (joining) gene segments that are separated in the germline and become assembled by site-specific recombination during early B and T cell differentiation (Tonegawa 1983). The rearrangements of V, D, and J segments are mediated by the products of two genes, *Rag-1* and *-2*, (Oettinger et al. 1990; Schatz and Baltimore 1988; Schatz et al. 1989). *Rag* expression is regulated both on the transcriptional as well as on the post-transcriptional level during the cell cycle and during lymphocyte development.

   Mice carrying a disrupted *Rag-1* or *Rag-2* gene are unable to initiate V(D)J recombination, which locks their Ig and TCR gene loci in germline configuration and, as a consequence, renders them unable to develop peripheral B and T lymphocytes (Mombaerts et al. 1992; Shinkai et al. 1992).

   Although V, D and J gene segments within the immunoglobulin and the T cell receptor loci share identical structures of recombination signal sequences (RSS) flanking the segments, Ig gene rearrangements are only completed in B cells and TCR rearrangements only in the T lineage indicating that *Rag*

[1] Basel Institute for Immunology, Grenzacherstr. 487, 4005 Basel, Switzerland
[2] Present address: Division of Molecular Oncology, Departments of Pathology, Medicine and Biochemistry, Washington University School of Medicine, 660 South Euclid Avenue, St. Louis, MO 63119, USA

expression alone is not sufficient to mediate rearrangement of endogenous Ig and TCR gene loci. Rearrangements of endogenous Ig or TCR gene loci are further regulated by transcription factors that are expressed stage- and tissue-specifically, which act on various enhancers within Ig and TCR gene loci (FERRIER et al. 1989; YANCOPOULOS and ALT 1986). Rearrangements within the Ig and TCR genes occur usually in an ordered fashion, with DJ (on IgH or TCR α or δ) rearrangements preceding V to DJ rearrangements, which in turn are completed before the IgL or TCR α and γ gene segments are assembled (ALLISON 1993; ALT et al. 1984, 1986; KRUISBEEK 1993). These stepwise rearrangements are preceded by an activation of transcription of the corresponding gene loci. If either the transcription factors or the enhancer elements involved in the regulation of this transcription are deleted by targeted mutation, rearrangements are inhibited at the various stages (BAIN et al. 1994; DORSHKIND 1994; HAGMAN and GROSSCHEDL 1994; SERWE and SABLITZKY 1993; TAKEDA et al. 1993; URBÁNEK et al. 1994; ZOU et al. 1993). This suggests that the transcription of the Ig or TCR gene locus is a prerequisite for its capacity to be rearranged.

We first give an overview of the different stages of B cell development in mouse bone marrow. We then describe how the expression of the *Rag*–1 and *Rag*–2 genes changes during this development on the mRNA and protein level. Finally, we present evidence that RAG-1 and RAG-2 protein influence each other so that, in the absence of RAG-2, the half life of RAG-1 is prolonged.

2 Regulation of *Rag* Gene Expression During Early Mouse B Cell Differentiation

2.1 Early B Cell Differentiation in Mouse Bone Marrow

B lymphopoiesis in adult mice occurs in the bone marrow and follows a program of differentiation that includes successive rearrangements on first the Ig heavy and later the Ig light chain gene loci (ALT et al. 1984). The expression of Ig proteins expressed from productively rearranged Ig gene loci is crucial for the selection of developing lymphocytes into various preB- and B cell compartments and thus for the generation of normally sized precursor and mature B cell pools (for a recent review see MELCHERS et al. 1995). Stages of early B cell differentiation can therefore be defined on the basis of the rearrangement status of their endogenous Ig gene loci. Cells that are committed to the B-lineage, but carrying all their Ig gene loci in germline configuration are designated progenitor (pro-) B cells (Fig. 1). Precursor B cells of type I (preB-I cells) are generated from these cells if at least one DJ-rearrangement has occurred on either of the two heavy chain alleles. Further V to DJ rearrangement leads to the stages of preB-II cells that can differentiate into immature, surface IgM bearing B cells after completion of a productive L chain

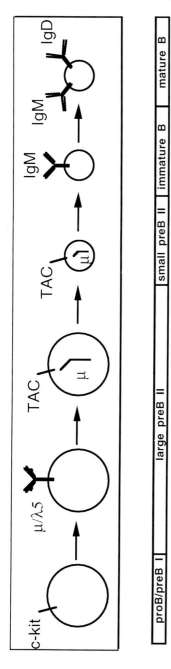

Rearrangments

	proB/preB I	large preB II	small preB II	immature B	mature B
heavy chain:	DJ/germline DJ/DJ, DJ/VDJ-	VDJ+/DJ VDJ+/VDJ-	VDJ+/DJ VDJ+/VDJ-	VDJ+/DJ VDJ+/VDJ-	VDJ+/DJ VDJ+/VDJ-
light chain:	germline	germline	germline	VJ	VJ	VJ
Cells in S/G2+M	35%	70%	60-70%	<5%	<2%	<2%

Fig. 1. Schematic representation of early B cell differentiation stages including surface markers and the possible Ig gene rearrangement status on both alleles. The cell surface markers used for sorting of B220⁺ cells are displayed in the scheme

gene rearrangement. Selection into the peripheral, mature lymphocyte pool changes these cells to longer-lived antigen-reactive cells which coexpress IgM and IgD on their surface.

All early B cell differentiation stages, from proB cells up to the antigen responsive B cells are characterized by the expression of the pan-B cell markers B220 (or CD45R) (COFFMAN 1982) and CD19 (ROLINK et al. 1996), which are expressed on 40%–50% of all nucleated cells in the bone marrow (ROLINK and MELCHERS 1993). With additional surface markers, early B cell differentiation stages can further be subdivided into subpopulations that either correlate in large part with the rearrangement status of Ig H and/or L chain gene loci, or that can be discriminated in functional terms (ROLINK et al. 1994, 1995).

One of the earliest surface markers known to be co-expressed on $CD19^+/B220^+$ B-lineage cells is encoded by the proto-oncogene *c-kit* (OGAWA et al. 1991), which is also found on pluripotent stem cells (IKUTA and WEISSMAN 1992) and the earliest extrathymic T-lineage committed progenitor T cells (RODEWALD et al. 1994). Most $B220^+/c\text{-}kit^+$ cells in the bone marrow of wild type mice are DJ_H-rearranged preB-I cells (ROLINK et al. 1993b). However, in *Rag-1* or *Rag-2* deficient mice, which are unable to initiate V(D)J rearrangements, or in J_HT mice (CHEN et al. 1993; EHLICH et al. 1993) which are unable to rearrange the H chain loci $B220^+/c\text{-}kit^+$ cells are present in normal, or even slightly elevated numbers (GRAWUNDER et al. 1995b). These cells probably represent lineage committed proB cells. They are likely to be present as a small part of the $B220^+/c\text{-}kit^+$ compartment in wild type mice (ROLINK et al. 1993b). In young animals (3–6 weeks of age) the population of proB and preB-I cells accounts for 5%–10% of total $B220^+$ cells in the bone marrow, declining with age (ROLINK et al. 1993b).

Virtually all of these cells are clonable on stromal cells in the presence of interleukin-7 (IL-7). While proliferating, these cells keep their differentiation stage. Removal of IL-7 from the culture induce V_H to D_HJ_H rearrangements and a differentiation without proliferation to sIg^+ and sIg^- immature B cells (ROLINK et al. 1991).

In order to generate a large compartment of VDJ-rearranged preB-II cells in bone marrow, expression and membrane deposition of a µH chain is required. These preB-II cells usually account for 50%–60% of all $B220^+$ cells in normal mouse bone marrow (ROLINK and MELCHERS 1993) and are characterized by upregulation of surface expression of CD25 (the IL-2Rα chain) (CHEN et al. 1994a; ROLINK et al. 1994) and the downregulation of c-kit expression (ROLINK et al. 1994).

Early B cell development can further be characterized by the expression of the surrogate light chain encoded by the two preB cell-specific genes λ5 and V_{preB} (KUDO and MELCHERS 1987, SAKAGUCHI and MELCHERS 1986) (Fig. 1). ProB and PreB-I cells express the surrogate L chain in association with a complex of glycoproteins (KARASUYAMA et al. 1993). As soon as a productive V_HDJ_H rearrangement has occurred, the preB-II cell will express a µHC/surrogate light chain containing preB cell receptor on the surface (KARASUYAMA et al. 1990;

TSUBATA and RETH 1990). All further stages of development are enriched to over 95% for μheavy chain (μHC)-expressing cells, indicating that a positive selection of productively over nonproductively rearranged cells occurs as soon as μHC can be expressed.

In *Rag*-deficient as well as in J$_H$-T, μM-T and λ5-T mice, the preB-II compartment is reduced at least 20–50 fold (KITAMURA et al 1991, 1992; ROLINK et al. 1993c, 1994). This indicates that the expression of the preB receptor on the surface of preB-II cells induces a proliferative expansion phase which effects the normally sized preB-II compartment. In line with this picture of B cell development are observations that he preB-II compartments is filled up to normal size with cycling and resting cells, when a transgenic μH chain is expressed in the *Rag* deficient genetic background (SPANOPOULOU et al. 1994; YOUNG et al. 1994, ROLINK et al. 1994).

All these results suggest the following scenario for early B cell development. After D to J$_H$ rearrangements have ocurred in preB-I cells, usually on both alleles (ROLINK et al. 1991), the generation of a productive VDJ heavy chain rearrangement allows μHC expression and the formation of a preB cell receptor. The preB cell receptor signals proliferative expansion of in-frame heavy-chain rearranged preB-II cells. It also induces its own downregulation so that preB-II cells become resting and fill up the pool of small, resting B220$^+$/CD25$^+$ preB-II cells, in which L chain gene rearrangements are initiated (TEN BOEKEL et al. 1995). Since this population accounts for approximately 30%–35% of the total B220$^+$ cells (and approximately 3/4 of all B220$^+$/CD25$^+$ preB-II cells) in the bone marrow, sufficient immature B cells expressing membrane bound IgM can be generated from these cells without further cell division (ROLINK and MELCHERS 1993). The repertoire of immature B cells is negatively selected against autoreactive cells, since these cells apparently become arrested in their differentiation and deleted in situ (SPANOPOULOU et al. 1994).

2.2 Rag Gene Expression in Early B Cell Development

With a set of monoclonal antibodies specific for the different proB to B cell stages described above (see also Fig. 1), we have enriched cells at stages of early B cell differentiation i.e., from the pro/preB-I cell stage to the mature peripheral B cell stage by fluorescence-activated cell sorting (FACS) and have analyzed these populations for the expression of the *Rag* genes on mRNA and protein levels. In particular, a monoclonal antibody (SL156) that specifically recognizes the μH chain/surrogate L chain preB cell receptor complex (WINKLER et al. 1995) has facilitated the isolation of a subpopulation of preB-II cells which express the preB cell receptor immediately after functional heavy chain VDJ rearrangement.

Rag expression on protein as well as on mRNA level was found to be high in the earliest proB/preB-I cell subpopulation (GRAWUNDER et al. 1995a) (Fig. 2). This might be expected since most of heavy chain DJ as well as VDJ

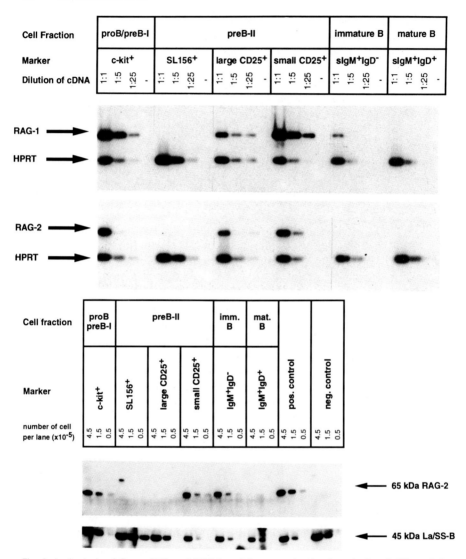

Fig. 2a,b. Analysis of *Rag* mRNA and RAG-2 protein expression in six early B cell differentiation subsets sorted by FACS from normal mouse bone marrow. **a** *Rag* mRNA expression compared with expression of the housekeeping gene HPRT (hypoxanthine-guanine phosphoribosyl transferase) analyzed by reverse-transcriptase-polymerase chain reaction (RT-PCR). **b** RAG-2 protein content compared to the ubiquitousely expressed nuclear SS-B protein as analyzed by western blot (Redrawn from GRAWUNDER et al. 1995a)

rearrangements appear to be carried out in this compartment. Analyses of *Rag*–1 and-2 mRNA and protein expression during the cell cycle of proliferating preB-I cells show that mRNA is expressed throughout the cycle. However, RAG-2 protein expression is reduced to levels below detection during the S, G2 and M phases of the cell cycle (GRAWUNDER et al. 1995a). This finding supports earlier studies by Lɪɴ and Dᴇsɪᴅᴇʀɪᴏ (1994) demonstrating that steady state levels of RAG-2 protein in cycling cells are low in S, G2 and M as a consequence of a reduced half-life of the protein.

As soon as one functional VDJ rearrangement has occurred on either of the two heavy chain alleles and the preB cell receptor is expressed on the surface, these cells downregulated *Rag-1* and *Rag-2* expression at least 50–100 fold, and RAG-2 protein expression was found to be below detection limits in Western blot analysis (Gʀᴀᴡᴜɴᴅᴇʀ et al. 1995a) (Fig. 2). This suggests that the shutdown of *Rag* expression after a functional VDJ rearrangement and preB receptor expression may, in fact, contribute to allelic exclusion of the Ig H chain loci.

In the proliferating in-frame heavy-chain rearranged preB-II cells, RAG-2 protein expression remains undetectable (in large, cycling $B220^+/CD25^+$ preB-II cells). However, *Rag* mRNA expression is detectable again in the subpopulation of cycling preB-II cells which no longer express the preB receptor. It is also notable that transcription of the κL chain locus is initiated although the L chain loci are not yet rearranged.

Rag mRNA expression in small $B220^+/CD25^+$ preB-II cells reaches levels again which are comparable to those found in the earliest $B220^+/c-kit^+$ pro/preB-I cells. The upregulation of *Rag* expression in these cells coincides with the rearrangement of the Ig L chain gene loci (ᴛᴇɴ Bᴏᴇᴋᴇʟ et al. 1995). RAG-2 protein is detectable at high levels in these resting preB-II cells. Finally, as found by others (Lɪ et al. 1993), *Rag* mRNA expression is downregulated as soon as IgM becomes expressed on the surface of immature B cells. However, RAG-2 protein expression remains high in this population (Fig. 2). This might be because the transition from small $CD25^+$ preB-II cells to immature B cells occurs without further cell division and because RAG-2 protein might acquire a longer half-life. As expected, *Rag* expression is no longer detectable in mature surface IgM^{low}, IgD^+ mature B cells (Fig. 2).

3 Expression of *Rag* Genes in the Lymphoid Lineage

Expression of both *Rag-1* and *Rag-2* genes in mammals is only detectable at significant levels in bone marrow and thymus (Oᴇᴛᴛɪɴɢᴇʀ et al. 1990; Sᴄʜᴀᴛᴢ et al. 1989). Further subfractionation of cellular differentiation stages within the B and T lineage has demonstrated that *Rag* mRNA expression is in fact confined to the earliest antigen receptor negative progenitor and precursor B and T cell stages ($B220^+IgM^-$ cells in the bone marrow and CD4⁻CD8⁻TCR- DN, as well as $CD4^+CD8^+$ DP cells in the thymus) (Lɪ et al. 1993; Tᴜʀᴋᴀ et al. 1991; Wɪʟsᴏɴ et al. 1994). *Rag* mRNA expression has been found by others to be markedly downregulated or even undetectable after surface deposition of surface IgM or TCR expression in $B220^+IgM^+IgD^-$ immature B cells of the bone marrow or in CD4+CD8- and CD4-CD8+ single positive thymocytes, respectively (Li et al. 1993; Turka et al. 1991). The downregulation of *Rag* mRNA observed after antigen receptor expression in immature B and T cells can be simulated by treatment of precursor B or T cells with phorbolesters, thereby mimicking the signalling function of surface IgM or TCR (Neale et al. 1992; Turka et al. 1991). Downregulation of Rag expression can also be achieved by crosslinking of surface IgM on transformed immature B cell lines from Eμ-N-myc transgenic mice that still express *Rag* mRNA (Ma et al. 1992).

Whereas crosslinking of antigen receptors in immature B cell lines in vitro (Ma et al. 1992) or during positive selection of TCR in the thymus (Brandle al. 1994) results in downregulation of *Rag* expression, maintained *Rag* expression has been observed when immature B cells receive a crosslinking receptor signal in the bone marrow (Tiegs et al. 1993). This process may enable a replacement of a given receptor with autoreactivity (Tiegs et al. 1993), and therefore has been named "receptor editing" (Gay et al. 1993; Radic et al. 1993; Tiegs et al. 1993).

In T cell development in the thymus, the regulated expression pattern of *Rag* genes was analyzed by Wilson and co-workers (Wilson et al. 1994). Two waves of *Rag* gene expression were seen. The first wave of expression appeared to coincide with β chain gene rearrangements in CD25$^+$ DN thymocytes, the latter with the onset of α gene rearrangements in TCR$^-$ DP thymocytes. In between these two waves a downregulation of *Rag* mRNA was observed. It is this stage of development where the TCR-β chain presumably associates with the pre-Tα chain (Saint-Ruf et al. 1994) on early thymocytes prior to TCR α chain gene rearrangements (Groettrup and von Boehmer 1993). Targeted disruption of the pre-Tα chain negatively affects development of α,β T cells in a similar way as the targeted disruption of the λ5 gene affects B cell development (Kitamura et al. 1992, Fehling et al. 1995). The role of the pre-Tα chain for T lymphopoiesis appears to be equivalent to that of λ5 for B lymphopoiesis. In B cell as well as in T cell development *Rag* gene expression and recombination activity pauses after a functional rearrangement of the heavy chain respectively TCR β chain while a clonal expansion of in-frame rearranged precursors takes place.

4 *Rag* Expression in B Lymphocyte Precursor Cell Lines In Vitro

Precursor B cell lines from bone marrow or fetal liver can either be established by transformation with Abelson murine leukaemia virus (A-MuLV) or grown as long-term proliferating untransformed cell lines by maintaining them on a stromal cells in the presence of IL-7 (Rolink et al. 1991). *Rag* gene expression in A-MuLV transformed preB cell lines is generally low, when compared with expression levels found, e.g., in the bone marrow (RATHBUN et al. 1993). In fact, A-MuLV transformed cell lines with higher passage numbers usually displayed the lowest *Rag* expression levels (ALT et al. 1992). Similar results were found with stromal cell/IL-7 dependent preB cells that express *Rag* mRNA levels comparable to those found in B220$^+$/c-kit$^+$ preB-I cells shortly after removal from the organ, but gradually downregulate *Rag* mRNA expression upon prolonged time in tissue culture (T.H. WINKLER and U. GRAWUNDER, unpublished observations).

However, unlike A-MuLV transformed preB cells, stromal cell/IL-7 dependent preB cells can be induced to differentiate in vitro, which can be achieved by the withdrawal of IL-7. This in vitro differentiation includes the upregulation of *Rag-1* and *Rag-2* expression to levels found in B220$^+$/c-kit$^+$ or small B220$^+$/CD25$^+$ preB cells (GRAWUNDER et al. 1995b). Furthermore, the cells become small resting lymphocytes, acquire a more mature phenotype (characterized, e.g., by the gain of expression of CD25 and the loss of expression of other preB cell specific markers), and they rearrange their endogenous IgH and L chain gene loci. PreB cells from wild-type mice also enter into a program of apoptosis. Apoptosis can be prevented, if cells from Eμ-bcl-2 transgenic mice (STRASSER et al. 1990) are used for this assay (ROLINK et al. 1993a). Apart from the prevention of apoptosis, the in vitro differentiation of Eμ-bcl-2 transgenic cells is identical to those of wild-type cells, including the upregulation of *Rag* expression (ROLINK et al. 1993a).

B lineage differentiation in vitro similar to that found with the stromal cell/Il-7 reactive preB-I cells is observed with conditionally A-MuLV transformed preB cell lines (ENGELMAN and ROSENBERG 1987). They can be grown at a permissive temperature of 34° C at a preB-I stage. When those cells are transferred to 39° C, the mutant virus becomes non-permissive resulting in growth arrest, the induction of an apoptosis program, the upregulation of *Rag* gene expression and the onset of κL chain gene rearrangements like in in vitro differentiating preB-II cells (CHEN et al. 1994b; KLUG et al. 1994).

The inducibility of endogenous *Rag* mRNA expression in either the temperature-sensitive (ts) A-MuLV transformed or the stromal cell/IL-7 dependent preB cell lines suggests that they are ideal tools for the identification and characterization of the V(D)J recombinase activity in a cell-free system. Indeed, the first report, describing the initiation of V(D)J recombination in a cell-free system, has taken advantage of ts A-MuLV transformed preB cell lines (VAN GENT et al. 1995). However, the efficiency of the reaction was very low in extracts alone and had to be supplemented with exogenously added recombinant RAG-1 protein.

Analysis of RAG-protein expression upon induction of stromal cell/IL-7 dependent preB cells has revealed that although both *Rag-1* and *Rag 2* mRNA expression become equally induced within 2–3 days, only RAG-2 protein accumulated with the same kinetic as its mRNA, whereas RAG-1 protein expression was only initially (within 18–24 h) induced, but dropped to very low expression levels thereafter (U. GRAWUNDER, P. SHOCKETT, T.J.M. LEU, A. ROLINK,

D.G. Schatz, and F. Melchers, manuscript in preparation). In contrast, in the absence of RAG-2 protein expression, i.e., in proB cells from RAG-2-T mice, RAG-1 protein accumulated to high steady state levels, with a kinetic resembling the induction of *Rag-1* mRNA. This finding suggests that the turnover of RAG-1 protein is influenced by the expression of RAG-2 protein. The nature of this crosstalk between the two RAG proteins is currently under investigation.

It can be assumed that this kind of regulation of RAG protein expression might also apply to ts A-MuLV transformed preB cells and that the limiting levels of RAG-1 protein in nuclear extracts from temperature induced preB cells therefore had to be supplemented with exogenously added recombinant RAG-1 protein, in order to increase the efficiency of the reaction (van Gent et al. 1995).

5 Conclusion

Rag expression in lymphocytes appears to be tightly regulated at both the transcriptional as well as the post-transcriptional level. It is tempting to speculate that this is of physiologic significance. It is conceivable that cells actively undergoing mitosis should not have high recombinase activity that involves DNA double strand breaks and therefore incurs the risk of recombining or losing genetic information. This requirement seems to be reflected by the fact that RAG-2 protein stability, and thus expression, is lowest at $S/G_2/M$ phases of the cell cycle.

It may further be disadvantageous if a lymphocyte expresses antigen receptors with two different specificities, since activation of a cell by one of its receptors, may also activate a response of the second one. Allelic exclusion of either IgH or TCR β chain gene loci might be in part effected by the shutdown of *Rag* expression at the transcriptional as well as the posttranslational levels.

Finally, it is conceivable that high expression levels of both RAG proteins might result in a high frequency of double strand breaks and recombination events to chromosomal regions other than the Ig and TCR gene loci. Chromosomal translocations to Ig and TCR loci, often the cause of B and T cell neoplasias, are evidence of the occurrence of such illegitimate V(D)J recombination events. This danger might be reduced if cells are able to respond to an overshooting V(D)J recombinase activity by limiting RAG-1 protein levels at a stage of development when both *Rag* genes are heavily transcribed.

Acknowledgements. The Basel Institute for Immunology was founded and is supported by the F. Hoffmann-La Roche, Basel Switzerland. U.G. is a post-doctoral fellow of the Boehringer Ingelheim Foundation, Stuttgart, Germany.

References

Allison JP (1993) γ,δ T-cell development. Curr Opin Immunol 5:241–246

Alt F, Yancopoulos GD, Blackwell TK, Wood C, Thomas E, Boss M, Coffman R, Rosenberg N, Tonegawa S, Baltimore D (1984) Ordered rearrangement of immunoglobulin heavy chain variable region segments. EMBO J 3:1209–1219

Alt FW, Blackwell TK, DePinho RA, Reth MG, Yancopoulos GD (1986) Regulation of genome rearrangement events during lymphocyte differentiation. Immunol Rev 89:5–30

Alt FW, Rathbun G, Oltz E, Taccioli G, Shinkai Y (1992) Function and control of recombination-activating gene activity. Ann NY Acad Sci 651:277–294

Bain G, Maandag ECR, Izon DJ, Amsen D, Kruisbeek AM, Weintraub BC, Krop I, Schlissel MS, Feeney AJ, van Roon M, van der Valk M, te Riele HPJ, Berns A, Murre C (1994) E2A proteins are required for proper B cell development and initiation of immunoglobulin gene rearrangements. Cell 79:885–892

Brandle D, Mller S, Mller C, Hengartner H, Pircher H (1994) Regulation of Rag–1 and CD69 expression in the thymus during positive and negative selection. Eur J Immunol 24:145–151

Chen J, Trounstine M, Alt FW, Young F, Kurahara C, Loring JF, Huszar D (1993) Immunoglobulin gene rearrangement in B cell deficient mice generated by targeted deletion of the J$_H$ locus. Int Immunol 5:647–656

Chen J, Ma A, Young F, Alt FW (1994a) IL-2 receptor α chain expression during early B lymphocyte differentiation. Int Immunol 6:1265–1268

Chen YY, Wang LC, Huang MS, Rosenberg N (1994b) An active v-abl protein tyrosine kinase blocks immunoglobulin light-chain gene rearrangement. Genes Dev 8:688–697

Coffman B (1982) Surface antigen expression and immunoglobulin rearrangement during mouse pre-B cell development. Immunol Rev 69:5–23

Dorshkind K (1994) Transcriptional control points during lymphopoiesis. Cell 79:751–753

Ehlich A, Schaal S, Gu H, Kitamura D, Mller W, Rajewsky K (1993) Immunoglobulin heavy and light chain genes rearrange independently at early stages of B cell development. Cell 72:695–704

Engelman A, Rosenberg A (1987) Isolation of temperature sensitive Abelson virus mutants by site-directed mutagenisis. Proc Natl Acad Sci USA 84:8021–8025

Fehling HJ, Krotkova A, Saint-Ruf C, von Boehmer H (1995) Crucial role of the pre-T-cell receptor a gene in development of αβ but not γδ T cells. Nature 375:795–798

Ferrier P, Krippl B, Furley AJ, Blackwell TK, Suh H, Mendelsohn M, Winoto A, Cook WD, Hood L, Costantini F, Alt FW (1989) Control of VDJ recombinase activity. Cold Spring Harbor Symposium on Quantitative Biology LIV:191–202

Gay D, Saunders T, Camper S, Weigert M (1993) Receptor editing: an approach by autoreactive B cells to escape tolerance. J Exp Med 177:999–1008

Grawunder U, Leu TMJ, Schatz DG, Werner A, Rolink AG, Melchers F, Winkler TH (1995a) Shutdown of expression of Rag–1 and Rag–2 in preB cells after functional immunoglobulin μ-heavy chain rearrangement. Immunity3:601–608

Grawunder U, Melchers F, Rolink A (1995b) Induction of sterile transcription from the κL chain gene locus in V(D)J-recombinase deficient progeniitor B cells. Int Immunol 8(12)

Groettrup M, von Boehmer H (1993) A role for a pre-T cell receptor in T cell development. Immunol Today 14:610–614

Hagman J, Grosschedl R (1994) Regulation of gene expression at early stages of B-cell differentiation. Curr Opin Immunol 6:222–230

Ikuta K, Weissman IL (1992) Evidence that mouse hematopoietic stem cells express c-kit but do not depend on steel factor for their generation. Proc Natl Acad Sci USA 89:1502–1506

Karasuyama H, Kudo A, Melchers F (1990) The proteins encoded by the V$_{preB}$ and λ$_5$ preB cell specific genes can associate with each other and with μ heavy chain. J Exp Med 172:969–972

Karasuyama H, Rolink A, Melchers F (1993) A complex of glycoproteins is associated with VpreB/lambda 5 surrogate light chain on the surface of μ heavy chain-negative early precursor B cell lines. J Exp Med 178:469–78

Kitamura D, Roes J, Khn R, Rajewsky K (1991) A B cell-deficient mouse by targeted disruption of the membrane exon of the immunoglobulin μ chain gene. Nature 350:423–426

Kitamura D, Kudo A, Schaal S, Mller W, Melchers F, Rajewsky K (1992) A critical role of λ$_5$ in B cell development. Cell 69:823–831

Klug CA, Gerety SJ, Shah PC, Chen YY, Rice NR, Rosenberg N, Singh H (1994) The *v-abl* tyrosine kinase negatively regulates NF-κB/Rel factors and blocks κ gene transcription in pre-B lymphocytes. Genes Dev 8:678–687

Kruisbeek AM (1993) Development of αβ T cells. Curr Opin Immunol 5:227–234

Kudo A, Melchers F (1987) A second gene, V_{preB} in the λ_5 locus of the mouse, which appears to be selectively expressed in preB lymphocytes. EMBO J 6:2267–2272

Li YS, Hayakawa K, Hardy RR (1993) The regulated expression of B lineage associated genes during B cell differentiation in bone marrow and fetal liver. J Exp Med 178:951–960

Lin WC, Desiderio S (1994) Cell cycle regulation of V(D)J recombination-activating protein RAG-2. Proc Natl Acad Sci USA 91:2733–2737

Ma A, Fisher P, Dildrop R, Oltz E, Rathbun G, Achacoso P, Stall A, Alt FW (1992) Surface IgM mediated regulation of *Rag* gene expression in *Eμ-N-myc* B cell lines. EMBO J 11:2727–2734

Melchers F, Rolink A, Grawunder U, Winkler TH, Karasuyama H, Ghia P, Andersson J (1995) Positive and negative selection events during B lymphopoiesis. Curr Opin Immunol 7:214–227

Mombaerts P, Iacomini J, Johnson RS, Herrup K, Tonegawa S, Papaioannou VE (1992) RAG-1-deficient mice have no mature B and T lymphocytes. Cell 68:869–877

Neale GA, Fitzgerald TJ, Goorha RM (1992) Expression of the V(D)J recombinase gene *Rag*-1 is tightly regulated and involves both transcriptional and post-transcriptional controls. Mol Immunol 29:1457–1466

Oettinger MA, Schatz DG, Gorka C, Baltimore D (1990) *Rag*-1 and *Rag*-2, adjacent genes that synergistically activate V(D)J recombination. Science 248:1517–1523

Ogawa M, Matzusaki Y, Nishikawa S, Hayashi SI, Kunisada T, Sudo T, Kina T, Nakauchi H, Nishikawa SI (1991) Expression and function of c-kit in hemopoietic progenitor cells. J Exp Med 174:63–70

Radic MZ, Erikson J, Litwin S, Weigert M (1993) B lymphocytes may escape tolerance by revising their antigen receptors. J Exp Med 177:1165–1173

Rathbun G, Oltz EM, Alt FW (1993) Comparison of *Rag* gene expression in normal and transformed precursor lymphocytes. Int Immunol 5:997–1000

Rodewald HR, Kretschmar K, Takeda S, Hohl C, Dessing M (1994) Identification of pro-thymocytes in murine fetal blood:T lineage commitment can precede thymus colonization. EMBO J 13:4229–4240

Rolink A, Melchers F (1993) Generation and regeneration of cells of the B-lymphocyte lineage. Curr Opin Immunol 5:207–217

Rolink A, Kudo A, Karasuyama H, Kikuchi Y, Melchers F (1991) Long-term proliferating early preB cell lines and clones with the potential to develop to surface-Ig positive mitogen-reactive B cells "in vitro" and "in vivo". EMBO J 10:327–336

Rolink A, Grawunder U, Haasner D, Strasser A, Melchers F (1993a) Immature surface Ig$^+$ B cells can continue to rearrange κ and λ L chain gene loci. J Exp Med 178:1263–1270

Rolink A, Haasner D, Nishikawa SI, Melchers F (1993b) Changes in frequencies of clonable preB cells during life in different lymphoid organs of mice. Blood 81:2290–2300

Rolink A, Karasuyama H, Grawunder U, Haasner D, Kudo A, Melchers F (1993c) B cell development in mice with a defective λ_5 gene. Eur J Immunol 23:1284–1288

Rolink A, Grawunder U, Winkler TH, Karasuyama H, Melchers F (1994) IL-2 receptor α chain (CD25,TAC) expression defines a crucial stage in preB cell development. Int Immunol 6:1257–1264

Rolink A, Andersson J, Ghia P, Grawunder U, Haasner D, Karasuyama H, ten Boekel E, Winkler TH, Melchers F (1995) B-cell development in mouse and man. Immunologist 3/4:125–128

Rolink A, ten Boekel E, Melchers F, Fearon DT, Krop I, Andersson J (1996) A subpopulation of B220$^+$ cells in murine bone marrow does not express CD19 and contains NK cell progenitors. J Exp Med 183:187–194

Saint-Ruf C, Ungewiss K, Groettrup M, Bruno L, Fehling HJ, von Boehmer H (1994) Analysis and expression of a cloned pre-T cell receptor gene. Science 266:1208–1212

Sakaguchi N, Melchers F (1986) λ_5, a new light-chain-related locus selectively expressed in preB lymphocytes. Nature 324:579–582

Schatz DG, Baltimore D (1988) Stable expression of immunoglobulin gene V(D)J recombinase activity by gene transfer into 3T3 fibroblasts. Cell 53:107–115

Schatz DG, Oettinger MA, Baltimore D (1989) The V(D)J recombination activating gene, *Rag*-1. Cell 59:1035–1048

Serwe M, Sablitzky F (1993) V(D)J recombination in B cells is impaired but not blocked by targeted deletion of the imunoglobulin heavy chain intron enhancer. EMBO J 12:2321–2327

Shinkai Y, Rathbun G, Lam KP, Oltz EM, Stewart V, Mendelsohn M, Charron J, Datta M, Young F, Stall AM, Alt FW (1992) RAG-2-deficient mice lack mature lymphocytes owing to inability to initiate V(D)J rearrangement. Cell 68:855–867

Spanopoulou E, Roman CAJ, Corcoran LM, Schlissel MS, Silver DP, Nemazee D, Nussenzweig MC, Shinton SA, Hardy RR, Baltimore D (1994) Functional immunoglobulin transgenes guide ordered B-cell differentiation in RAG-1 deficient mice. Genes Dev 8:1030–1042

Strasser A, Harris AW, Vaux DL, Webb E, Bath ML, Adams JM, Cory S (1990) Abnormalities of the immune system induced by dysregulated bcl-2 expression in transgenic mice. Curr Top Microbiol Immunol 166:175–181

Takeda S, Zou YR, Bluethmann H, Kitamura D, Muller U, Rajewsky K (1993) Deletion of the immunoglobulin κ chain intron enhancer abolishes κ chain gene rearrangements in cis but not λ chain rearrangements in trans. EMBO J 12:2329–2336

ten Boekel E, Melchers F, Rolink A (1995) The status of Ig loci rearrangements in single cells from different stages of B cell development. Int Immunol 7:1013–1019

Tiegs SL, Russell DM, Nemazee D (1993) Receptor editing in self-reactive bone marrow B cells. J Exp Med 177:1009–1020

Tonegawa S (1983) Somatic generation of antibody diversity. Nature 302:575–581

Tsubata T, Reth M (1990) The products of preB cell specific genes ($λ_5$ and V_{preB}) and the immunoglobulin μ chain form a complex that is transported onto the cell surface. J Exp Med 172:973–976

Turka LA, Schatz DG, Oettinger MA, Chun JJ, Gorka C, Lee K, McCormack WT, Thompson CB (1991) Thymocyte expression of RAG-1 and RAG-2:termination by T cell receptor cross-linking. Science 253:778–781

Urbánek P, Wang ZQ, Fetka I, Wagner EF, Busslinger M (1994) Complete block of early B cell differentiation and altered patterning of the posterior midbrain in mice lacking Pax5/BSAP. Cell 79:901–912

van Gent DC, McBlane JF, Ramsden DA, Sadofsky MJ, Hesse JE, Gellert M (1995) Initiation of V(D)J recombination in a cell-free system. Cell 81:925–934

Wilson A, Held W, MacDonald HR (1994) Two waves of recombinase gene expression in developing thymocytes. J Exp Med 179:1355–1360

Winkler TH, Rolink A, Melchers F, Karasuyama H (1995) Precursor B cells of mouse bone marrow express two different complexes with surrogate light chain on the surface. Eur J Immunol 25:446–450

Yancopoulos GD, Alt FW (1986) Regulation of the assembly and expression of variable-region genes. Ann Rev Immunol 4:339–368

Young F, Ardman B, Shinkai Y, Lansford R, Blackwell TK, Mendelsohn M, Rolink A, Melchers F, Alt FW (1994) Influence of immunoglobulin heavy- and light-chain expression on B-cell differentiation. Genes Dev 8:1043–1057

Zou YR, Takeda S, Rajewsky K (1993) Gene targeting in the Igκ locus:efficient generation of λ chain-expressing B cells, independent of gene rearrangements in Igκ. EMBO J 12:811–820

The Cell Cycle and V(D)J Recombination

Stephen Desiderio[1], Weei-Chin Lin[2], and Zhong Li[1]

1 Introduction

V(D)J recombination, the process by which lymphocyte antigen receptor genes are assembled during lymphocyte development, is the only known form of site-specific DNA rearrangement in vertebrates (Lewis 1994). V(D)J recombination is initiated by double-strand DNA cleavage at conserved recombinational signal sequences (Roth et al. 1992a,b,1993; Schlissel et al. 1993; McBlane et al. 1995; vanGent et al. 1995), in this way resembling other DNA transactions such as yeast mating-type switch recombination (Herskowitz 1989). The involvement of double-strand cleavage in these processes is potentially deleterious, since introduction or persistence of double-strand breaks at recombinational signal sequences during M phase or S phase could interfere with faithful transmission of genetic information to daughter cells. Eukaryotic cells have evolved protective checkpoint mechanisms that sense the presence of double-strand DNA breaks and induce delays in cell cycle progression from G1 into S or from G2 into M. These delays mitigate against attempted replication of damaged templates and missegregation of damaged chromosomes.

---

[1] Department of Molecular Biology and Genetics and Howard Hughes Medical Institute, The Johns Hopkins University School of Medicine, 725 North Wolfe Street Baltimore, Maryland 21210, USA
[2] Department of Medicine, Duke University Medical Center, Durham, North Carolina 27710, USA

Mutations that eliminate delays at these checkpoints are associated with increased sensitivity to ionizing radiation and genetic instability. For example, in patients with the autosomal recessive disorder ataxia-telangiectasia the G1 checkpoint is impaired (KASTAN et al. 1992); this fundamental defect is accompanied by increases in radiosensitivity and in the incidence of malignancies, particularly lymphoid tumors (KASTAN 1995). Circulating lymphocytes from patients with ataxia-telangiectasia exhibit a several hundredfold increase in the frequency of interlocus V(D)J recombination (LIPKOWITZ et al. 1990), suggesting that enforcement of the G1 cell-cycle checkpoint enhances the fidelity of V(D)J recombination.

For physiologic processes such as V(D)J recombination, protection against such harmful effects could be reinforced by regulatory mechanisms that prohibit double-strand DNA cleavage during the S and M cell cycle phases. Indeed, in the case of yeast mating-type switch recombination, the initiating DNA cleavage is accomplished by an endonuclease, HO, whose expression is restricted to the G1 phase of cell cycle (NASMYTH 1985). Initiation of V(D)J recombination could also be controlled by regulating site-specific DNA cleavage. This could be accomplished by controlling expression or activity of the recombination activating proteins RAG-1 or RAG-2, which together are necessary and sufficient for site-specific double-strand cleavage at V(D)J recombination signals (MCBLANE et al. 1995; VANGENT et al. 1995). Indeed, available evidence indicates that RAG activity – and consequently, initiation of V(D)J recombination – is controlled at the transcriptional and post-transcriptional levels.

2 Double-Strand DNA Cleavage and Double-Strand Break Repair in V(D)J Recombination

V(D)J recombination is mediated by conserved heptamer and nonamer signal sequences, separated by less highly conserved spacer regions of 12 or 23 bp. The recombination signal sequences are oriented so that heptamer elements abut coding segments. V(D)J recombination is initiated by site-specific, double-stranded DNA cleavage at the junctions between coding sequences and heptamer recombinational signals (Fig. 1). The apparent products of these site-specific cleavage reactions can be found in thymus and in transformed B lymphoid cells (ROTH et al. 1992a,b, 1993; SCHLISSEL et al. 1993) and are of two types. The signal ends are predominantly blunt, phosphorylated at their 5' ends, and terminate in an intact heptamer element. The coding ends, unlike signal ends, terminate in hairpin structures; the occurrence of palindromic DNA insertions (P elements) at coding junctions supports the interpretation that the coding ends detected in these experiments are V(D)J recombination intermediates. Similar products are generated in vitro by cleavage with purified RAG-1 and RAG-2 proteins (MCBLANE et al. 1995; VANGENT et al. 1995).

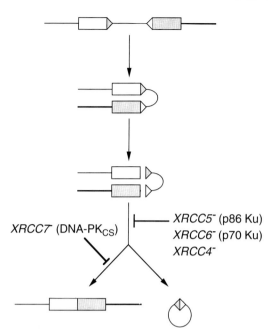

Fig. 1. Dependence of V(D)J recombination on double-strand break repair genes. Two antigen receptor gene segments are shown; coding sequences are represented by rectangles, recombination signal sequences by triangles. Site-specific DNA cleavage by the RAG-1 and RAG-2 proteins produces an intermediate containing four DNA ends; the coding ends terminate in hairpins, while the signal ends are blunt. Mutations in four different double-strand DNA repair genes exhibit impairments in V(D)J recombination. Mutations in *XRCC7*, corresponding to the mouse *scid* locus, selectively impair coding joint formation. Mutations in *XRCC4*, *XRCC5* and *XRCC6* impair formation of coding and signal joints. *XRCC7* encodes the catalytic subunit of a 450-kDa, DNA-dependent protein kinase (DNA-PK$_{CS}$). *XRCC5* encodes the p86 subunit of Ku antigen, a heterodimeric DNA end-binding protein that associates with DNA-PK$_{CS}$

The cleaved coding and signal ends are subsequently joined in what appear to be distinct reactions (Ramsden and Gellert 1995). Joining of coding sequences is frequently accompanied by modification of nucleotide sequences at recombinant junctions (through insertion of P elements, nucleotide loss, or non-templated nucleotide addition). In contrast, joining of signal sequences generally involves precise fusion of the heptamer elements (Lewis et al. 1985). These outcomes are not exclusive; less frequently, the recombination signal sequences of one gene segment may be fused intact to the coding sequence of another gene segment, or gene segments may be opened and resealed at the junction between the coding sequence and the conserved heptamer (Lewis et al. 1988; Morzycka-Wroblewska et al. 1988).

V(D)J rearrangement requires the RAG-1 and RAG-2 proteins, whose genes were originally identified by their ability to activate rearrangement of an exogenous V(D)J recombinational substrate in fibroblastoid cells (Schatz et al.

1989; OETTINGER et al. 1990). In the intact mouse, RAG-1 and RAG-2 are both needed for development of functional B and T cells (MOMBAERTS et al. 1992; SHINKAI et al. 1992). Increased RAG expression in vivo is accompanied by an increase in the frequency of V(D)J recombination (MENETSKI and GELLERT 1990; OLTZ et al. 1993); moreover, accumulation of double-stranded breaks at recombination signal sequences is dependent on RAG expression (SCHLISSEL et al. 1993), indicating a requirement for the RAG proteins at an early stage in V(D)J rearrangement. The recent development of a cell-free system that carries out site-specific, recombination signal-dependent DNA cleavage in vitro has allowed identification of RAG-1 and RAG-2 as the sole essential components of the endonuclease that inititates V(D)J recombination (McBLANE et al. 1995; vanGENT et al. 1995). Whether the RAG proteins also play a role in the joining of cleaved recombination intermediates is not known.

The joining of coding and signal ends is clearly dependent on several proteins that function more generally in double-strand DNA break repair. Mutations in genes corresponding to four mammalian complementation groups, originally associated with hypersensitivity to ionizing radiation, are also deficient in V(D)J recombination (TACCIOLI et al. 1993; PERGOLA et al. 1993; LEE et al. 1995; ZDZIENICKA 1995) (Fig. 1). The products of three of these genes – XRCC5, XRCC6, and XRCC7 – have been identified; the product of the fourth gene, XRCC4, has yet to be described. Mutations in XRCC4, XRCC5, and XRCC6 impair formation of coding and signal joints (TACCIOLI et al. 1993; PERGOLA et al. 1993; LEE et al. 1995). Mutations in XRCC7, which corresponds to the mouse scid locus (TACCIOLI et al. 1994a), affect coding joint formation selectively (BLACKWELL et al. 1989; MALYNN et al. 1988; LIEBER et al. 1988). As a result of impaired V(D)J recombination, homozygous scid mice are deficient in functional B and T cells (BOSMA et al. 1983); scid cells exhibit hypersensitivity to gamma irradiation and a general defect in double-strand DNA break repair (BIEDERMANN et al. 1991; HENDRICKSON et al. 1991; FULOP and PHILLIPS 1990). XRCC7 encodes the catalytic subunit of a 460-kDa DNA-dependent protein kinase, DNA-PK$_{CS}$ (PETERSON et al. 1995; KIRCHGESSNER et al. 1995; BLUNT et al. 1995; HARTLEY et al. 1995). XRCC5 encodes the larger subunit of Ku antigen, a p86/p70 heterodimer that binds DNA ends and associates with DNA-PK$_{CS}$ (GETTS and STAMATO 1994; SMIDER et al. 1994; TACCIOLI et al. 1994b; RATHMELL and CHU 1994); XRCC6 mutants are also defective in Ku activity, suggesting that XRCC6 encodes the smaller Ku subunit (BOUBNOV and WEAVER 1995). DNA-PK$_{CS}$ and several other DNA damage-response proteins, including the product of the ataxia-telangiectasia gene (SAVITSKY et al. 1995), share significant amino acid sequence homology to phosphotidylinositol (PI) 3-kinases, although DNA-PK$_{CS}$ appears to function exclusively as a protein kinase (HARTLEY et al. 1995). The ataxia-telangiectasia product and the complex of Ku and DNA-PK$_{CS}$ may function in signal transduction pathways that trigger cellular responses to DNA damage. The dependence of V(D)J recombination on proteins that function in double-strand break repair suggests that these two DNA transactions are coordinately regulated.

3 Regulation of *RAG* Transcripts by Extrinsic Factors

The *RAG-1* and *RAG-2* genes are transcribed exclusively in immature lymphoid cells, where their expression is regulated in response to developmental cues. Antigen receptors, cytokine receptors and costimulatory molecules have all been shown to regulate *RAG* expression at the transcriptional level. Steady-state levels of *RAG* transcripts fall coordinately upon crosslinking of antigen receptors on nascent lymphocytes (TURKA et al. 1991; MA et al. 1992). During thymic selection, *RAG* transcripts are extinguished upon engagement of the T-cell receptor (TCR) by major histocompatibility (MHC) molecules; this effect is reversed by anti-MHC class I antibodies (BRANDLE et al. 1994). Recent evidence indicates that expression of *RAG* transcripts may be regulated reciprocally by interleukin-7 (IL-7) and CD19 during B cell development. In B cell progenitors from human bone marrow treatment with IL-7 was followed by increased expression of CD19 and later by diminished expression of *RAG-1* and *RAG-2*; IL-7-induced suppression of *RAG* transcripts was abrogated by crosslinking of CD19 (BILLIPS et al. 1995). The responsiveness of *RAG* expression to engagement of a diverse set of receptors provides rich opportunities for regulation of V(D)J recombination during lymphocyte development.

4 Periodic Accumulation of RAG-2 Protein in Cycling Lymphoid Cells

Restriction of V(D)J recombination to a particular cell cycle interval might require finer temporal control of RAG expression or activity than can be provided by transcriptional regulation alone. This requirement could be met by one or more post-transcriptional mechanisms, including regulation of RAG protein synthesis or degradation, sequestration of one or both RAG proteins, and modulation of RAG activity by post-translational modification. In dividing lymphoid cells, accumulation of RAG-2 protein is tightly regulated in the cell cycle, oscillating as a function of cell cycle phase by a factor of 20 or more (LIN and DESIDERIO 1994). RAG-2 protein accumulates preferentially in G1; the amount of RAG-2 diminishes sharply as cells enter S phase and remains low or undetectable until after M phase. By contrast, the amount of RAG-1 protein is invariant across cell cycle. Cell cycle periodicity of RAG-2 protein accumulation is not accompanied by variation in the steady-state level of *RAG-2* RNA, and is therefore established at the post-transcriptional level. Because the rate of RAG-2 protein synthesis, as assessed by pulse labeling experiments, is similar in the G1 and G2/M cell cycle phases, regulation of RAG-2 protein degradation appears to be principally responsible for the periodic accumulation of RAG-2 in cycling cells (Z. LI and S. DESIDERIO, unpublished results).

5 Coupling of RAG-2 Protein Degradation to the Cell Cycle

One means of tethering intracellular processes to the cell cycle clock is through periodic phosphorylation by one or more cyclin-dependent kinases. The cyclin-dependent kinase $p34^{cdc2}$ phosphorylates RAG-2 in vitro at a single predominant site, Thr490 (LIN and DESIDERIO 1993). In transfected fibroblastoid cells the half-life of wild-type RAG-2 is only approximately 12 min; mutation of Thr490 to alanine (T490A) is associated with a 20-fold increase in RAG-2 half-life (about 200 min) and a corresponding increase in its accumulation (LIN and DESIDERIO 1993). The prolonging effect of the T490A mutation on RAG-2 half-life is independent of RAG-1 expression. These observations suggested that RAG-2 is targeted for rapid degradation by phosphorylation at Thr490.

Consistent with this idea, phosphorylation of Thr490 in vivo was detectable only when the degradation of RAG-2 protein was suppressed (LIN and DESIDERIO 1993). Chimeric proteins containing a 90-amino acid segment of RAG-2 spanning Thr490 or the T490A mutation exhibit differences in steady-state accumulation similar to that observed between intact wild-type and mutant RAG-2. A comparison of RAG-2 sequences from xenopus, chicken, rabbit, mouse, and humans reveals that this region is highly conserved phylogenetically (SADOFSKY et al. 1994), despite the fact that it is dispensable for V(D)J rearrangement (SADOFSKY et al. 1994; SILVER et al. 1993; CUOMO and OETTINGER 1994). In particular, a minimal consensus motif for phosphorylation by cyclin-dependent kinases (corresponding to Thr-Pro at residues 490 and 491 of mouse RAG-2) is strictly conserved throughout phylogeny. Taken together, these observations suggest that phosphorylation of Thr490 by one or more cyclin-dependent kinases is required to generate a local structural signal for the degradation of RAG-2 protein.

This idea is further supported by systematic mutational analysis of the 90-amino acid C-terminal region of mouse RAG-2. These experiments reveal that rapid degradation of RAG-2 protein requires two essential elements: (1) the invariant Thr-Pro residues comprising a cyclin-dependent kinase (cdk) phosphorylation site at positions 490 and 491; and (2) a group of cationic residues that reside within a 10-amino acid interval downstream of the phosphorylation site (Z. LI and S. DESIDERIO, unpublished results). A RAG-2 mutant in which this region (amino acid residues 499–508) is replaced by alanine residues (RAG-2 499–508A) accumulates to a level similar to that observed for the T490A and P491A mutant proteins (Z. LI and S. DESIDERIO, unpublished results). The synthetic rates of RAG-2 T490A, RAG-2 P491A and RAG-2 499–508A are similar;

pulse chase experiments reveal that the half-lives of the mutant proteins are correlated with their steady-state amounts relative to wild-type RAG-2 (Z. LI and S. DESIDERIO, unpublished results). It has yet to be determined whether the 499–508A mutation affects the same degradation pathway as mutations in the cdk phosphorylation site. It is also important to note that a link between phosphorylation and degradation of RAG-2 has not yet been demonstrated directly, and it remains possible that mutations at residues 490 and 491 prolong the half-life of RAG-2 by means other than interference with phosphorylation.

Strikingly, the RAG-2 degradation signal also governs the periodic disappearance of RAG-2 protein in cycling cells. The cell cycle-dependence of RAG-2 expression was examined in fibroblastoid cells expressing wild-type or mutant RAG-2 proteins by stable transfection (Z. LI and S. DESIDERIO, unpublished results). Cells were removed from cycle by serum withdrawal and then induced to divide synchronously by serum supplementation. Accumulation of wild-type RAG-2 protein oscillated as cells traversed the cell cycle, declining to undetectable levels upon entry into S phase and recovering as cells reentered G1. In contrast, RAG-2 proteins carrying mutations at the cdk phosphorylation site or within the 499–508 interval were expressed at invariant levels throughout the cell cycle. These observations are consistent with a model in which the periodic accumulation of RAG-2 protein in dividing cells is governed by phosphorylation and dephosphorylation of RAG-2 at Thr490 (Fig. 2). In this model, phosphorylation marks RAG-2 for rapid degradation during the S, G2 and M

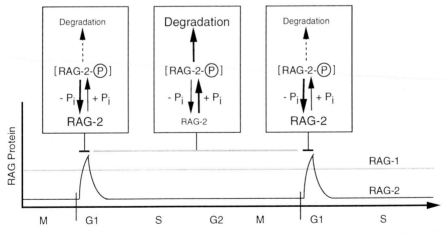

Fig. 2. A model for regulation of RAG-2 accumulation in cycling cells. Amounts of RAG-1 and RAG-2 proteins in a cycling cell are shown schematically as a function of cell cycle phase. The positions of the curves are not intended to indicate the actual relative levels of RAG-1 and RAG-2. The amount of RAG-1 protein remains relatively constant throughout the cell cycle, while RAG-2 is expressed periodically. According to this model: (1) RAG-2 is targeted for rapid degradation near the G1-S boundary upon phosphorylation at Thr490 by one or more cyclin-dependent kinases; (2) the Thr490-phosphorylated form of RAG-2 predominates through S, G2 and M, and RAG-2 remains unstable; and (3) dephosphorylation of RAG-2 during G1 leads to prolongation of RAG-2 protein half-life and increased accumulation

cell cycle phases; dephosphorylation during G1 would remove the mark for degradation and allow accumulation of RAG-2.

6 Evidence for Proteasomal Degradation of RAG-2

ATP-dependent, ubiquitin-dependent proteolysis is implicated in the selective, rapid degradation of many proteins, including cyclin B, p53, c-*fos* and E1A (HERSHKO and CIECHANOVER 1992). In some instances (for example, in the case of ornithine decarboxylase) degradation is ATP-dependent but ubiquitin-independent (BERCOVICH et al. 1989). In both degradation pathways proteolysis is carried out by high molecular weight (26S or 20S) proteasomal complexes. Recent experiments employing the pharmacologic inhibitors N-acetyl-Leu-Leu-norleucinal (LLnL) or N-acetyl-Leu-Leu-methioninal (LLM) suggest that one or more proteasomal degradation pathways can participate in degradation of RAG-2 (Z. LI and S. DESIDERIO, unpublished results). Within several hours of treatment of transfected cells with LLnL, a substantial increase in accumulation of RAG-2 protein was observed; the effect was specific, as the less potent compound LLM had little effect on RAG-2 accumulation (Z. LI and S. DESIDERIO, unpublished results). Although these observations suggest that RAG-2 degradation involves one or more proteasomal complexes, they do not indicate the extent to which proteasomal degradation might contribute to the rapid turnover of RAG-2 outside of G1.

7 Preferential Accumulation of RAG-2 in Resting Intrathymic T-Cell Progenitors

Examination of RAG expression in normal thymocytes is complicated by developmental heterogeneity and by the presence of noncycling (G0) cells. In mouse (LIN and DESIDERIO 1994) or avian (FERGUSON et al. 1994) thymocytes, RAG-2 protein is expressed preferentially in G0/G1 phase cells and is expressed at low or undetectable levels in cells traversing the S or G2/M (LIN and DESIDERIO 1994). Unlike RAG-2, RAG-1 protein persists during the S and G2/M cell cycle phases (LIN and DESIDERIO 1994). The paucity of RAG-2 protein and the persistence of RAG-1 expression in S and G2/M phase thymocytes agree with patterns observed in lymphoid cell lines.

Levels of *RAG-1* and *RAG-2* RNA are several-fold greater in resting than in cycling thymocytes (W.-C. LIN and S. DESIDERIO, unpublished results). Developmental variation in the level of *RAG* RNA accumulation may contribute to this difference, as the resting thymocyte fraction is enriched in CD4$^+$CD8$^+$TCR^{lo}

cells, in which *RAG* RNA expression is elevated (see below). The elevation in *RAG* RNA observed in the resting fraction is commensurate with an increase in the level of RAG-1 protein. RAG-2, like RAG-1, accumulates preferentially in the resting thymocyte fraction. RNA accumulation does not, however, fully account for variation in the amount of RAG-2 protein; *RAG-2* transcripts persist in S- and G2/M-enriched fractions despite the absence of detectable protein product. One possible explanation is that in thymus the accumulation of RAG-1 protein is governed principally or exclusively at the level of transcription, while expression of RAG-2 protein is subject to both transcriptional and post-transcriptional control.

8 Accumulation of V(D)J-Associated Double-Strand DNA Breaks in G0/G1 Thymocytes

Taken together, the restriction of RAG-2 protein expression to G0/G1 thymocytes and the RAG-2-dependence of DNA cleavage at V(D)J recombination signal sequences are consistent with the observed distribution of V(D)J-associated double-strand breaks as a function of thymocyte DNA content. In thymocyte nuclei separated according to DNA content, or in intact thymocytes separated according to size, double-strand signal breaks at immunoglobulin (Ig) J_H segments (SCHLISSEL et al. 1993) or the TCR Dδ2 segment (Fig. 3) were found to accumulate preferentially in G0/G1 phase cells. Because accumulation of V(D)J-associated double-strand DNA breaks depends on their kinetics of generation and removal, these observations do not in themselves establish that cleavage at V(D)J recombination signal sequences occurs preferentially in G0/G1, but this interpretation is greatly strengthened by the observed periodicity of RAG-2 accumulation.

9 Physiologic Correlates

In development of both T- and B-cells, rearrangement of antigen receptor H-chain (TCR β or Ig H) and L-chain (TCR α or Ig L) genes occurs preferentially during two periods of quiescence or slow cellular division, separated by a period of rapid proliferation. For T-cell development the available data (GODFREY et al. 1993, 1994; BOYER et al. 1989; DUDLEY et al. 1994) indicate the following sequence: (1) after an early proliferative expansion of CD4⁻CD8⁻CD44^hiIL-2R⁺ intrathymic T-cell progenitors, thymocytes enter a nondividing or slowly dividing state (CD4⁻CD8⁻CD44^loIL-2R⁺) in which TCR β rearrangement occurs; (2) appearance of a functional TCR β chain, perhaps as part of a TCR β-surrogate

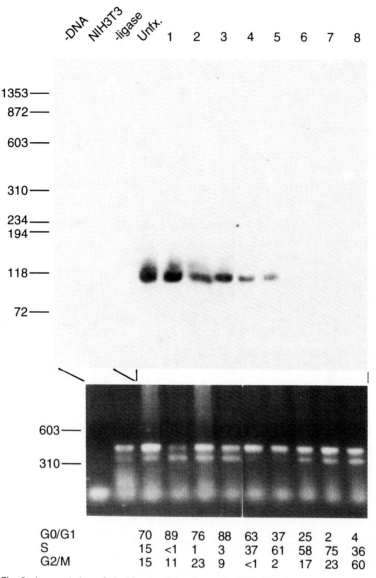

Fig. 3. Accumulation of double-strand breaks at the TCR Dδ2 locus as a function of cell cycle phase. Double-strand breaks at the TCR Dδ2 locus were assayed in elutriated mouse thymocytes by a ligation-mediated polymerase chain reaction (PCR) assay as described (ROTH et al. 1993). Thymocytes from 7-day-old mice were separated by centrifugal elutriation. An aliquot of each fraction was assayed for DNA content by flow cytometry. Genomic DNA (3 μg) from each fraction was used in the linker-mediated PCR assay for double-strand breaks at the TCR Dδ2 locus. Amplification of a signal end-containing molecule cleaved at the junction with the Dδ2 coding sequence is expected to yield a 119-bp fragment capable of hybridizing to an internal probe corresponding to sequences immediately upstream of the recombination signal nonamer. This product was readily detected in unfractionated thymocytes (*Unfx.*) and in elutriation fractions enriched for G0/G1 cells; the product was greatly diminished or undetectable in fractions enriched for cells in S or G2/M (*upper panel*). An internal control sequence, consisting of a portion of the *RAG-1* gene, was amplifiable in all fractions (*bottom panel*). For the unfractionated sample and for each elutriation fraction (*lanes 1–6*), the percentage of cells in G0/G1, S and G2/M are indicated below. The sizes of DNA standards, in basepairs, are indicated at left

α chain (gp33) complex (GROETTRUP et al. 1993; GROETTRUP and VON BOEHMER 1993; SAINTRUF et al. 1994), triggers proliferative expansion of a CD4⁻CD8⁻ CD44^{lo}IL-2R⁻ population; (3) this expansion is followed by the appearance of a smaller, nondividing CD4⁺CD8⁺ population, in which most TCR α rearrangement occurs.

A similar pattern is observed in B-cell development (KARASUYAMA et al. 1994; ROLINK et al. 1994): (1) assembly of Ig μ-chain genes occurs within a slowly dividing, CD45R⁺c-kit⁺IL-2R⁻CD43⁺ population (pro-B/pre-B-I); (2) productive Ig μ rearrangement is accompanied by surface expression of surrogate L-chain, acquisition of the CD45R⁺c-kit⁻IL-2R⁺ phenotype and initiation of rapid cycling (large pre-B-II); (3) large pre-B-II cells subsequently give rise to a resting, CD45R⁺c-kit⁻IL-2R⁺ population (small pre-B-II) which is the principal site of Ig L-chain gene rearrangement. Thus, in the development of both lymphoid lineages antigen receptor heavy-chain gene rearrangement is followed by rapid cell division and supression of V(D)J recombination; this period of cellular expansion is followed by a resting period in which most L-chain gene rearrangement occurs.

V(D)J recombination could be restricted to particular developmental periods by control at the level of the substrate or of the recombinational machinery. Substrate level regulation (e.g., locus accessibility) may be essential for timing of rearrangements during development and for allelic exclusion; there is yet no evidence suggesting a role for substrate level regulation in control of V(D)J recombination during the cell cycle. Two lines of evidence suggest that regulation of *RAG* gene expression contributes to control of V(D)J recombination during lymphocyte development. First, *RAG* transcripts accumulate preferentially at two discrete times during development, corresponding to the immature lymphocyte subsets in which H- and L-chain gene rearrangements are most active. In T-cell ontogeny these are the CD4⁻CD8⁻ CD44^{lo}CD25⁺ and CD4⁺CD8⁺TCR β⁺ subsets (WILSON et al. 1994); in B-cell ontogeny, the pro-B/pre-BI and small pre-BII subsets (KARASUYAMA et al. 1994; ROLINK et al. 1994). Second, as described above, accumulation of *RAG* transcripts can be modulated by extrinsic factors such as IL-7. In mixed culture systems containing stromal cells and B-cell progenitors, withdrawal of IL-7 is associated with growth arrest, increased accumulation of *RAG-1* and *RAG-2* RNA and initiation of Ig gene rearrangement (ROLINK et al. 1993).

Another link between V(D)J recombination and the cell cycle may be provided by post-transcriptional regulation of RAG-2 expression. Transitions from quiescence to rapid cycling, as observed in the progression from CD44^{lo}IL-2R⁺ to CD44^{lo}IL-2R⁻ or from pro-B/pre-B-I to large pre-B-II, are expected from observations summarized above to be accompanied by rapid degradation of RAG-2 protein at the G1-S transition. RAG-2 expression – and consequently, V(D)J recombination – might continue to be suppressed during rapid cell division if the G1 phase were sufficiently short. Such a mechanism could provide a pause in recombination during which the TCR β or Ig μ loci could be closed to the recombinase and the TCR α or Ig κ loci opened for subsequent rearrangement. Evidence consistent with this hypothesis was recently provided by the observation that large pre-B-II cells express *RAG-2* RNA but do not express detectable RAG-2 protein; RAG-2 protein levels become detectable only upon the transition from the large pre-B-II to the small (i.e., resting) pre-B-II stage (U. GRAWUNDER, T. WINKLER, and F. MELCHERS, personal communication).

10 Conclusions

The involvement of double-strand DNA cleavage in antigen receptor gene rearrangement first suggested the possibility that initiation of V(D)J recombination is regulated in the cell cycle. In developing lymphocytes, regulation of *RAG* transcription appears to contribute to developmental regulation of V(D)J recombination and to enhanced expression of RAG proteins in resting cells. Posttranscriptional control plays a prominent role in the regulation of RAG-2 expression in dividing cells, by restricting accumulation of RAG-2 protein – and by inference, initiation of V(D)J recombination – to the G1 cell cycle phase. The precise mechanism by which RAG-2 protein accumulates periodically in the cell cycle is not yet known, but may involve the triggering of protein degradation by specific phosphorylation. How phosphorylation might target RAG-2 for degradation and how RAG-2 is subsequently degraded are questions whose answers are expected to have interesting general implications. Cell cycle regulation of V(D)J recombination may serve to protect against deleterious effects of double-strand DNA cleavage, but other functions – for example, in enforcement of allelic exclusion – are also conceivable. These potential roles may be gauged in the future by examining the physiologic effects of unscheduled RAG expression in the mouse.

References

Bercovich Z, Rosenberg-Hasson, Y, Ciechanover, A, Kahana, C (1989) Degradation of ornithine decarboxylase in reticulocyte lysate is ATP-dependent but ubiquitin-independent. J Biol Chem 264:15949–15952

Biedermann KA, Sun J, Giaccia AJ, Tosto LM, Brown JM (1991) *scid* mutation in mice confers hypersensitivity to ionizing radiation and a deficiency in DNA double-strand break repair. Proc Natl Acad Sci USA 88:1394–1397

Billips LG, Nunez CA, Bertrand FE, Stankovic AK, Gartland GL, Burrows PD, Cooper MD (1995) Immunoglobulin recombinase gene activity is modulated reciprocally by interleukin 7 and CD19 in B cell progenitors. J Exp Med 182:973–982

Blackwell TK, Malynn BA, Pollock RR, Ferrier P, Covey LR, Fulop GM, Phillips RA, Yancopoulos GD, Alt FW (1989) Isolation of *scid* pre-B cells that rearrange kappa light chain genes: formation of normal signal and abnormal coding joins. EMBO J 8:735–742

Blunt T, Finnie N, Taccioli G, Smith G, Demengeot J, Gottlieb T, Mizuta R, Varghese A, Alt F, Jeggo P, Jackson S (1995) Defective DNA-dependent protein kinase activity is linked to V(D)J recombination and DNA repair defects associated with the murine *scid* mutation. Cell 80:813–823

Bosma GC, Custer RP, Bosma MJ (1983) A severe combined immunodeficiency mutation in the mouse. Nature 301:527–530

Boubnov NV, Weaver DT (1995) *scid* cells are deficient in Ku and replication protein A phosphorylation by the DNA-dependent protein kinase. Mol Cell Biol 15:5700–5706

Boyer PD, Diamond RA, Rothenberg EV (1989) Changes in the inducibility of IL-2 receptor alpha chain and T-cell receptor expression during thymocyte differentiation in the mouse. J Immunol 142:4121–4130

Brandle D, Muller S, Muller C, Hengartner H, Pircher H (1994) Regulation of RAG-1 and CD69 expression in the thymus during positive and negative selection. Eur J Immunol 24:145–151

Cuomo CA, Oettinger MA (1994) Analysis of regions of RAG-2 important for V(D)J recombination. Nucl Acids Res 22:1810–1814

Dudley EC, Petrie HT, Shah LM, Owen MJ, Hayday AC (1994) T cell receptor β chain gene rearrangement and selection during thymocyte development in adult mice. Immunity 1:83–93

Ferguson SE, Accavitti MA, Wang DD, Chen CL, Thompson CB (1994) Regulation of RAG-2 protein expression in avian thymocytes. Mol Cell Biol 14:7298–7305

Fulop GM, Phillips RA (1990) The scid mutation in mice causes a general defect in DNA repair. Nature 347:479–482

Getts RC, Stamato TD (1994) Absence of a Ku-like DNA end binding activity in the xrs double-strand DNA repair-deficient mutant. J Biol Chem 269:15981–15984

Godfrey DI, Kennedy J, Suda T, Zlotnik A (1993) A developmental pathway involving four phenotypically and functionally distinct subsets of CD3⁻CD4⁻CD8⁻ triple-negative adult mouse thymocytes defined by CD44 and CD25 expression. J Immunol 150:4244–4252

Godfrey DI, Kennedy J, Mombaerts P, Tonegawa S, Zlotnik A (1994) Onset of TCR-β rearrangement and role of TCR-β expression during CD3⁻CD4⁻CD8⁻ thymocyte differentiation. J Immunol 152:4783–4792

Groettrup M, von Boehmer H (1993) A role for a pre-T-cell receptor in T-cell development. Immunol Today 14:610–4

Groettrup M, Ungewiss K, Azogui O, Palacios R, Owen MJ, Hayday AC, von Boehmer H (1993) A novel disulfide-linked heterodimer on pre-T cells consists of the T cel receptor beta chain and a 33 kd glycoprotein. Cell 75:283–294

Hartley KO, Gall D, Smith GCM, Zhang H, Divecha N, Connelly MA, Admon A, Lees-Miller SP, Anderson CW,Jackson SP (1995) DNA-dependent protein kinase catalytic subunit: a relative of phosphatidylinositol 3-kinase and the ataxia telangiectasia gene product. Cell 82:849–856

Hendrickson EA, Qin X-Q, Bump EA, Schatz DG, Oettinger M, Weaver DT (1991) A link between double-strand bread-related repair and V(D)J recombination: the *scid* mutation. Proc Natl Acad Sci USA 88:4061–4065

Hershko A, Ciechanover A (1992) The ubiquitin system for protein degradation. Ann Rev Biochem 61:761–807

Herskowitz I (1989) A regulatory hierarchy for cell specialization in yeast. Nature 342:749–757

Karasuyama H, Rolink A, Shinkai Y, Young F, Alt FW, Melchers F (1994) The expression of Vpre-B/lambda 5 surrogate light chain in early bone marrow precursor B cells of normal and B cell-deficient mutant mice. Cell 77:133–143

Kastan M (1995) Ataxia-telangiectasia – broad implications for a rare disorder. N Engl J Med 333:662–663

Kastan MB, Zhan Q, El-Deiry WS, Carrier F, Jacks T, Walsh WV, Plunkett BS, Vogelstein B, Fornace AJ Jr (1992) A mammalian cell cycle checkpoint pathway utilizing p53 and GADD45 is defective in ataxia-telangiectasia. Cell 71:587–597

Kirchgessner C, Patil C, Evans J, Cuomo C, Fried L, Carter T, Oettinger M, Brown J (1995) DNA-dependent kinase (p350) as a candidate gene for the murine *scid* defect. Science 267:1178–1183

Lee SE, Pulaski CR, He DM, Benjamin DM, Voss M, Um J, Hendrickson EA (1995) Isolation of mammalian cell mutants that are X-ray sensitive, impaired in DNA double-strand break repair and defective for V(D)J recombination. Mutat Res 336:279–291

Lewis S (1994) The mechanism of V(D)J joining: lessons from molecular, immunological and comparative analyses. Adv Immunol 56:27–150

Lewis SA, Gifford A, Baltimore D (1985) DNA elements are asymmetrically joined during the site-specific recombination of kappa immunoglobulin genes. Science 228:677–685

Lewis SM, Hesse JE, Mizuuchi K, Gellert M (1988) Novel strand exchanges in V(D)J recombination. Cell 55:1099–1107

Lieber MR, Hesse JE, Lewis S, Bosma GC, Rosenberg N, Mizuuchi K, Bosma MJ, Gellert M (1988) The defect in murine severe combined immune deficiency: joining of signal sequences but not coding segments in V(D)J recombination. Cell 55:7–16

Lin W-C, Desiderio S (1993) Regulation of V(D)J recombination activator protein RAG-2 by phosphorylation. Science 260:953–959

Lin W-C, Desiderio S (1994) Cell cycle regulation of V(D)J recombination activating protein RAG-2. Proc Natl Acad Sci USA 91:2733–2737

Lipkowitz S, Stern MH, Kirsch IR (1990) Hybrid T cell receptor genes formed by interlocus recombination in normal and ataxia-telangiectasia lymphocytes. J Exp Med 172:409–418

Ma A, Fisher P, Dildrop R, Oltz E, Rathburn G, Achacoso P, Stall A, Alt FW (1992) Surface IgM mediated regulation of RAG gene expression in Eμ-N-myc B cell lines. EMBO J 11:2727–2734

Malynn BA, Blackwell TK, Fulop GM, Rathbun GA, Furley AJ, Ferrier P, Heinke LB, Phillips RA, Yancopoulos GD, Alt FW (1988) The scid defect affects the final step of the immunoglobulin VDJ recombinase mechanism. Cell 54:453–460

McBlane JF, vanGent DC, Ramsden DA, Romeo C, Cuomo CA, Gellert M, Oettinger MA (1995) Cleavage at a V(D)J recombination signal requires only RAG1 and RAG2 proteins and ooccurs in two steps. Cell 83:387–395

Menetski JP, Gellert M (1990) V(D)J recombination activity in lymphoid cell lines is increased by agents that elevate cAMP. Proc Natl Acad Sci USA 87:9324–9328

Mombaerts P, Iacomini J, Johnson RS, Herrup K, Tonegawa S, Papaisannou VE (1992) RAG-1-deficient mice have no mature B and T lymphocytes. Cell 68:869–877

Morzycka-Wroblewska E, Lee FEH, Desiderio SV (1988) Unusual immunoglobulin gene rearrangement leads to replacement of recombinational signal sequences. Science 242:261–263

Nasmyth K (1985) A repetitive DNA sequence that confers cell-cycle START (CDC28)-dependent transcription of the HO gene in yeast. Cell 42:225–235

Oettinger MA, Schatz DG, Gorka C, Baltimore D (1990) RAG-1 and RAG-2, adjacent genes that synergistically activate V9D)J recombination. Science 248:1517–1523

Oltz EM, Alt FW, Lin W-C, Chen H, Taccioli G, Desiderio S, Rathbun G (1993) A V(D)J recombinase-inducible B-cell line: role of transcriptional enhancer elements in directing V(D)J recombination. Mol Cell Biol 13:6223–6230

Pergola F, Zdzienicka MZ, Lieber MR (1993) V(D)J recombination in mammalian cell mutants defective in DNA double-strand break repair. Mol Cell Biol 13:3464–3471

Peterson SR, Kurimasa A, Oshimura M, Dynan, WS, Bradbury, EM, Chen, DJ (1995) Loss of the catalytic subunit of the DNA-dependent protein kinase in DNAA double-strand-break-repair mutant mammalian cells. Proc Natl Acad Sci USA 92:3171–3174

Ramsden DA, Gellert M (1995) Formation and resolution of double-strand break intermediates in V(D)J rearrangement. Genes Dev 9:2409–2420

Rathmell WK, Chu G (1994) Involvement of the Ku autoantigen in the cellular response to DNA double-strand breaks. Proc Natl Acad Sci USA 91:

Rolink A, Grawunder U, Haasner D, Strasser A, Melchers F (1993) Immature surface Ig+ B cells can continue to rearrange kappa and lambda L chain gene loci. J Exp Med 178:1263–1270

Rolink A, Grawunder U, Winkler TH, Karasuyama H, Melchers F (1994) IL-2 receptor alpha chain (CD25,Tac) expression defines a crucial stage in pre-B cell development. Int Immunol 6:1257–1264

Roth DB, Menetski JP, Nakajima PB, Bosma MJ, Gellert M (1992a) V(D)J recombination: broken DNA molecules with covalently sealed (hairpin) coding ends in scid mouse thymocytes. Cell 70:983–991

Roth DB, Nakajima PB, Menetski JP, Bosma MJ, Gellert M (1992b) V(D)J recombination in mouse thymocytes: double-strand breaks near T cell receptor δ rearrangement signals. Cell 69:41–53

Roth D, Zhu C, Gellert M (1993) Characterization of broken DNA molecules associated with V(D)J recombination. Proc Natl Acad Sci USA 90:10788–10792

Sadofsky MJ, Hesse JE, Gellert M (1994) Definition of a core region of RAG-2 that is functional in V(D)J recombination. Nucl Acids Res 22:1805–1809

Saintruf C, Ungewiss K, Groettrup M, Bruno L, Fehling HJ, Von Boehmer H (1994) Analysis and expression of a cloned pre-T cell receptor gene. Science 266:1208–1212

Savitsky K, Barshira A, Gilad S, Rotman G, Ziv Y, Vanagaite L, Tagle DA, Smith S, Uziel T, Sfez S, Ashkenazi M, Pecker I, Frydman M, Harnik R, Patanjali SR, Simmons A, Clines GA, Sartiel A, Gatti RA, Chessa L, Sanal O, Lavin MF, Jaspers NGJ, Malcolm A, Taylor R, Shiloh Y (1995) A single ataxia telangiectasia gene with a product similar to PI-3 kinase. Science 268:1749–1753

Schatz DG, Oettinger MA, Baltimore D (1989) The V(D)J recombination activating gene, RAG-1. Cell 59:1035–1048

Schlissel M, Constantinescu A, Morrow T, Baxter M, Peng A (1993) Double-strand signal sequence breaks in V(D)J recombination are blunt, 5'-phosphorylated, RAG-dependent, and cell cycle regulated. Genes Dev 72520–32:2520–2532

Shinkai Y, Rathbun G, Lam K-P, Oltz EM, Stewart V, Mendelsonn Y, Charron J, Datta M, Young F, Stall AM, Alt FW (1992) RAG-2-deficient mice lack mature lymphocytes owing to inability to initiate V(D)J rearrangement. Cell 68:855–867

Silver DP, Spanopoulou E, Mulligan RC, Baltimore D (1993) Dispensable sequence motifs in the RAG-1 and RAG-2 genes for plasmid V(D)J recombination. Proc Natl Acad Sci USA 90:6100–6104

Smider V, Rathmell WK, Lieber MR, Chu G (1994) Restoration of x-ray resistance and V(D)J recombination in mutant cells by Ku cDNA. Science 14:288–291

Taccioli GE, Rathbun G, Oltz E, Stamato T, Jeggo PA, Alt FW (1993) Impairment of V(D)J recombination in double-strand break repair mutants. Science 260:207–210

Taccioli GE, Cheng HL, Varghese AJ, Whitmore G, Alt FW (1994a) A DNA repair defect in Chinese hamster ovary cells affects V(D)J recombination similarly to the murine scid mutation. J Biol Chem 269:7439–7442

Taccioli GE, Gottlieb TM, Blunt T, Priestley A, Demengeot J, Mizuta R, Lehmann AR, Alt FW, Jackson SP, Jeggo PA (1994b) Ku80 – product of the XRCC5 gene and its role in DNA repair and V(D)J recombination. Science 265:1442–1445

Turka LA, Schatz DG, Oettinger MA, M CJJ, Gorda C, Lee K, McCormack WT, Thompson CB (1991) Thymocyte expression of RAG-1 and RAG-2: termination by T cell receptor cross-linking. Science 253:778–781

vanGent DC, McBlane JF, Ramsden DA, Sadofsky MJ, Hesse JE, Gellert M (1995) Initiation of V(D)J recombination in a cell-free system. Cell 81:925–934

Wilson A, Held W, Macdonald HR (1994) Two waves of recombinase gene expression in developing thymocytes. J Exp Med 179:1355–1360

Zdzienicka MZ (1995) Mammalian mutants defective in the response to ionizing radiation-induced DNA damage. Mutat Res 336:203–213

Double-Strand Breaks, DNA Hairpins, and the Mechanism of V(D)J Recombination

Sharri Bockheim Steen[1], Chengming Zhu[2], and David B. Roth[1,2]

1 Introduction

In vivo analysis of DNA intermediates has provided important mechanistic information about a number of recombination systems. DNA molecules containing double-strand breaks are well-established intermediates in several recombination systems and have been detected in vivo at hotspots for meiotic recombination (Cao et al. 1990; Sun et al. 1989) and during mating type switching in the yeast *Saccharomyces cerevisiae* (Connolly et al. 1988; Raveh et al. 1989; White and Haber 1990). Site-specific double-strand breaks have also been observed during the movement of transposable elements, such as Tc1 and Tc3 in *Caenorhabditis elegans* (van Luenen et al. 1993; Ruan and Emmons 1994) and Tn7 and Tn10 in bacteria (Bainton et al. 1991; Haniford et al. 1991). Characterization of these broken DNA molecules has afforded a better understanding of the sequence of events leading to product formation. In particular, analysis of the effects of mutations in *trans*-acting factors on processing of in vivo intermediates has been instrumental in defining reaction pathways (reviewed by Haber 1992). In the past several years, a similar analysis of double-strand breaks associated with V(D)J recombination has been undertaken. Although there is as yet no direct evidence that the broken molecules are intermediates, these studies, which are summarized below, have provided valu-

---

[1] Cell and Molecular Biology Program, Baylor College of Medicine, 1 Baylor Plaza, Houston, TX 77030, USA

[2] Department of Microbiology and Immunology, Baylor College of Medicine, 1 Baylor Plaza, Houston, TX 77030, USA

able information about the recombination mechanism and may guide the interpretation of more detailed biochemical studies in cell-free systems.

2 Signal Ends

One early model for V(D)J recombination (Fig. 1) suggested that recombination might be initiated by the introduction of double-strand breaks precisely between the recombination signal sequences (RSS) and the adjacent coding segments (ALT and BALTIMORE 1982). As shown in Fig. 1, a pair of RSS, one with a heptamer-12 base pair spacer-nonamer configuration and one with a heptamer-23 base pair spacer-nonamer configuration, are recognized by the recombination machinery. The presence of a 12 and a 23 base pair spacer RSS is required for efficient recombination, a reaction feature termed the "12/23 rule". According to the model, double-strand breaks are produced precisely at the border between each RSS heptamer and its adjacent coding segment, thus generating two types of DNA termini: coding ends and signal ends. The joining of coding ends produces a coding joint; signal ends join to form a signal joint. These two types of junctions differ in one important respect: signal joints almost never exhibit loss of nucleotides, whereas coding joints have frequently lost several nucleotides from one or both ends and often have gained extra nucleotides (reviewed by LEWIS 1994a). This asymmetry provided the first clue that signal and coding ends might be processed differently. Subsequent characterization of double-strand breaks generated in vivo has provided more direct evidence to support this hypothesis, as described below.

The double-strand cleavage model shown in Fig. 1 predicts that broken DNA ends, if sufficiently long-lived, might accumulate to detectable levels in cells that are very active for recombination of particular gene segments. Accordingly, initial attempts to characterize intermediates in V(D)J recombination focused on a particular T cell receptor (TCR) locus, TCRδ, that actively recombines in thymocytes of neonatal mice. Highly sensitive Southern blotting techniques were used to detect double-strand breaks at this locus in genomic DNA preparations from thymocytes of newborn mice (ROTH et al. 1992b). The most extensively characterized of these broken molecules (species with signal ends at the TCR Dδ2 element) is present in roughly 2% of thymocytes of newborn BALB/c mice. Other hybridization probes were used to detect molecules with signal ends resulting from cleavage at D and J coding elements. Interestingly, all broken molecules detected could be accounted for by cleavage at pairs of RSS; species resulting from cleavage at only a single RSS were not observed (ROTH et al. 1992b). These data suggest that cleavage may occur in a coordinated fashion at pairs of signals, as discussed below. Excised circular products containing signal joints were detected at about the same abundance

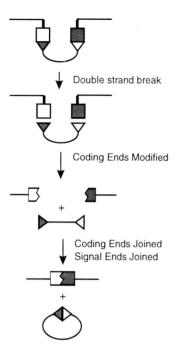

Fig. 1. A double-strand break model for V(D)J recombination. Recombination is initiated by recognition of a 12-spacer recombination signal sequence (RSS) (*open triangle*) and a 23-spacer RSS (*filled triangle*). The recombination machinery introduces a double-strand break precisely between the RSS and the coding segments (shown as *squares*), generating signal ends that join to form a product containing a signal joint and coding ends that, after possible modification (nucleotide loss and/or addition), are joined to form a coding joint

Double strand break

Coding Ends Modified

Coding Ends Joined
Signal Ends Joined

as broken molecules. The vast majority of the signal joints in these excision products are perfect (no loss or addition of nucleotides), as judged by sensitivity to a restriction endonuclease whose recognition site is created by precise joining of a pair of RSS (ROTH et al. 1992b).

Although the broken molecules identified in these studies were regarded as plausible reaction intermediates, the possibility that the broken species might represent "dead end" products could not be excluded. Recent in vivo time course experiments performed in the Gellert laboratory have made use of a cell line in which recombination activity can be induced. After induction, signal ends accumulate to high levels and can be apparently "chased" into joined products (RAMSDEN and GELLERT 1995). These results provide additional evidence suggesting that the broken molecules are in fact recombination intermediates.

Several methods have been used to characterize the broken ends in more detail. The structure of signal ends was probed by testing the ability of T4 DNA ligase to join a nonphosphorylated, double-stranded oligonucleotide with a blunt end to the signal ends. Ligation was assessed by a change in the size of the broken molecules as judged by Southern blotting. The majority of the signal ends were able to ligate to the oligonucleotide, indicating that most of the signal ends are blunt, unblocked (free of covalent modifications), and 5′ phosphorylated (ROTH et al. 1993). Signal ends were also examined using a much more sensitive and versatile technique based on ligation-mediated polymerase chain reaction (LMPCR). Although originally used to detect single-strand nicks (MUELLER and WOLD 1989), the assay was modified to specifically detect blunt ends by ligating a double-stranded oligonucleotide ligation primer that contains a blunt end to broken DNA molecules recovered from actively recombining cells (Fig. 2) (ROTH et al. 1993; SCHLISSEL et al. 1993). Since the ligation primers do not contain 5' phosphoryl groups, only molecules with blunt ends terminating in 5' phosphates are capable of ligating to the primer. PCR amplification is performed using a gene-specific primer and a primer that hybridizes to the ligated oligonucleotide. In addition to its high sensitivity, the LMPCR assay is semiquantitative (ROTH et al. 1993; ZHU and ROTH 1995; SCHLISSEL et al. 1993) and allows the structure of broken ends to be examined in detail. The precise site of cleavage can be obtained by conventional cloning and sequencing of PCR products (SCHLISSEL et al. 1993). Alternatively, the bulk population of amplified molecules can be examined by designing ligation primers that generate a unique restriction site upon ligation to a perfect signal end, so the status of the ends can be tested by simply digesting the PCR products

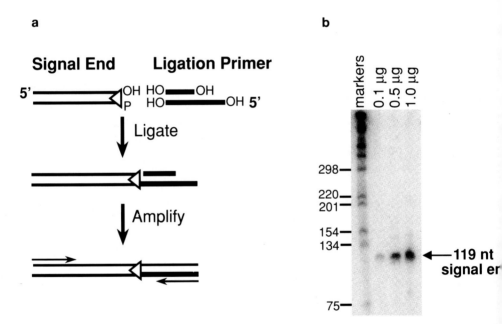

Fig. 2a,b. Detection of signal ends by ligation-mediated polymerase chain reaction (LMPCR). A simplified schematic of the assay is shown in **a**. LMPCR involves ligation of a double-stranded oligonucleotide (ligation primer) to a broken end, followed by PCR amplification using a gene-specific primer (*arrow on the left*) and a primer that binds to the ligated oligonucleotide (*arrow on the right*). Because the ligation primers are not phosphorylated, the 5' terminus of the broken end must contain a 5' phosphoryl group. Amplified products are separated by polyacrylamide gel electrophoresis and visualized by hybridization to a radiolabeled oligonucleotide probe. Detection of signal ends at the T cell receptor (TCR) Dδ2 locus in thymocyte DNA preparations from wild-type mice is shown in **b**. The indicated amounts of thymocyte DNA from neonatal BALB/c mice were ligated with the ligation primer and subjected to LMPCR analysis. Using this technique, Dδ2 signal ends can be detected in as little as 0.01 μg of thymocyte DNA. For details, see (ZHU and ROTH 1995)

with the appropriate restriction enzyme (ROTH et al. 1993). Data from both approaches indicate that the vast majority of signal ends retain the entire nucleotide sequence of the RSS, without loss or addition of nucleotides (SCHLISSEL et al. 1993; ROTH et al. 1993). Experiments in which genomic DNA was treated with various enzymes (phosphatase, polynucleotide kinase, Klenow fragment, T4 DNA polymerase) prior to LMPCR confirmed that the majority of signal ends are blunt and contain 5' phosphoryl groups (SCHLISSEL et al. 1993; ROTH et al. 1993). These data strongly suggest that signal ends are produced by precise cleavage between the RSS and the coding segment, in agreement with the double-strand cleavage model. The lack of modification of signal ends is consistent with the extremely low frequency of nucleotide loss observed at signal joints (LIEBER et al. 1988b).

Despite the presence of relatively abundant signal ends created by double-strand cleavage, the coding ends that are presumably created in this same reaction have not yet been detected in lymphoid cells of wild-type mice either by Southern blotting (ROTH et al. 1992b) or by much more sensitive LMPCR techniques (ROTH et al. 1993; SCHLISSEL et al. 1993; ZHU and ROTH 1995). The simplest explanation, that coding ends might be joined much more rapidly than signal ends, is supported by the recent observation that coding ends resulting from immunoglobulin κ gene rearrangement in cultured cell lines derived from wild-type mice appear to join much more rapidly than signal ends (RAMSDEN and GELLERT 1995). These results, along with the previously mentioned differences in junction structures, suggest that different steps may be involved in joining these two types of ends. This hypothesis is further supported by the differential effects of mutations in *trans*-acting factors on coding and signal joint formation, as discussed in the following section.

3 Coding Ends Accumulate in Thymocytes of scid Mice

Mice homozygous for a mutation (BOSMA et al. 1983) resulting in severe combined immune deficiency (scid mice) are incapable of producing significant levels of mature B and T lymphocytes due to a severe deficiency in V(D)J recombination (LIEBER et al. 1988a; MALYNN et al. 1988; HENDRICKSON et al. 1988). Southern blotting analysis of both endogenous antigen receptor loci in lymphocytes from scid mice (SCHULER et al. 1986) and integrated recombination substrates in scid cell lines (HENDRICKSON et al. 1988; MALYNN et al. 1988) revealed large deletions, presumably caused by faulty recombination. These data suggest that recombination is initiated in scid cells but that later steps, such as end processing or joining, proceed aberrantly. Experiments using extrachromosomal recombination substrates demonstrated that scid cell lines are severely defective in coding joint formation, while signal joints form relatively normally (Lieber et al. 1988a). This selective effect on coding joint formation suggests that the *scid* factor is required for proper joining of coding ends.

The *scid* defect also affects general repair of double-strand breaks in both lymphoid and nonlymphoid cells (FULOP and PHILLIPS 1990; HENDRICKSON et al. 1991b; BIEDERMANN et al. 1991). This observation provided the first clue that factors involved in double-strand break repair also participate in joining coding ends. Southern blotting analysis revealed that coding ends accumulate to high levels in scid thymocytes (ROTH et al. 1992a), providing the most direct evidence for impaired joining of coding ends. This is not simply due to lack of general end joining activity, as linear DNA molecules transfected into scid cell lines are joined normally (HARRINGTON et al. 1992; CHANG et al. 1993; LEWIS 1994b).

Why do coding ends accumulate in scid cells? One explanation was provided by the observation that all coding ends detected by Southern blotting in thymocytes of scid mice are covalently sealed in the form of DNA hairpins, as assessed by their resistance to exonuclease treatment and their migration pattern when subjected to two-dimensional native-alkaline gel electrophoresis (ROTH et al. 1992a). These coding ends were characterized in greater detail using a modified LMPCR assay in which hairpin ends are opened using a single-strand specific nuclease, allowing them to be detected by LMPCR (Fig. 3). These studies demonstrated that all detectable coding ends produced by cleavage at TCR Dδ2 in scid thymocytes terminate in hairpins (ZHU and ROTH 1995). No coding ends (hairpin or nonhairpin) were detected in wild-type cells, indicating that the abundance of coding ends is at least 1000-fold lower than signal ends in thymocytes of wild-type mice (ZHU and ROTH 1995). Examination of signal ends isolated from scid thymocytes revealed that both the abundance (ROTH et al. 1992a) and the structure (ZHU and ROTH 1995) of signal ends are not measurably affected by *scid* mutation. Thus, the *scid* defect results in accumulation of a very specific class of DNA molecules: covalently sealed coding ends.

4 Hairpin Coding Ends As Normal Intermediates in V(D)J Recombination

Three lines of evidence suggest that covalently sealed coding ends are normal intermediates in V(D)J recombination. First, hairpin coding ends resulting from immunoglobulin light chain gene rearrangement have been detected in cultured cell lines derived from wild-type mice (RAMSDEN and GELLERT 1995). Second, nuclear extracts from wild-type cells are capable of cleaving RSS-containing substrates, forming blunt signal ends and hairpin coding ends (VAN GENT et al. 1995). Third, a particular class of extra nucleotides, termed P nucleotides, is frequently present at coding joints produced in wild-type cells. These insertions differ from the nontemplated bases, termed N nucleotides, added by terminal

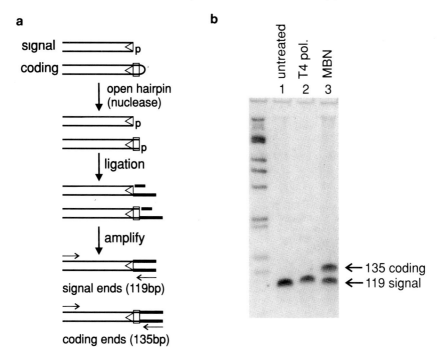

Fig. 3a,b. A ligation-mediated polymerase chain reaction (LMPCR)-based assay for detection of hairpin coding ends. A schematic of the assay used to detect hairpin coding ends is shown in **a**. To make hairpin ends available for ligation, DNA preparations are first treated with a single-strand specific endonuclease, such as mung bean nuclease or P1 nuclease. This treatment opens the hairpins, producing a population of molecules containing blunt-ended forms that can be joined to the ligation primers. These molecules are then amplified by PCR and detected as described in Fig. 2. Representative data are shown in **b**. Each lane contains PCR products derived from 1 μg of DNA prepared from thymocytes of scid mice. Prior to LMPCR, the samples were treated as follows: lane 1, no treatment; lane 2, pretreatment with T4 DNA polymerase to convert any nonblunt ends into flush termini capable of joining to the ligation primer; lane 3, pretreatment with mung bean nuclease (MBN) to open hairpins. See (ZHU and ROTH 1995) for details

deoxynucleotidyl transferase (TdT) (GILFILLAN et al. 1993; KOMORI et al. 1993; LIEBER et al. 1988b) in that P nucleotides are complementary to the last few nucleotides of the coding segment, forming a short palindrome (LAFAILLE et al. 1989; McCORMACK et al. 1989; MEIER and LEWIS 1993). These curious insertions therefore appear to be derived from the coding sequence itself and can be simply explained by opening hairpin coding ends in a particular fashion, as described below.

A scheme illustrating the production of P nucleotides from hairpin coding ends is shown in Fig. 4. According to this proposal, cleavage at the RSS generates a pair of blunt signal ends and a pair of covalently sealed coding ends. Hairpins are opened by introduction of a nick a few nucleotides away from the terminal base pair (termed the hairpin "tip"), leaving a short single-

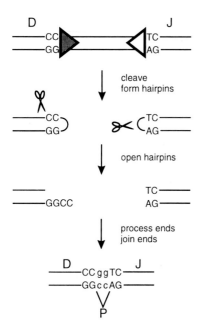

Fig. 4. A hairpin model for V(D)J recombination and P nucleotide formation. Coding ends, formed by precise cleavage between the signal and coding elements, exist in a covalently sealed "hairpin" conformation. Hairpin coding ends must presumably be opened prior to joining. Introduction of a nick near, but not at, the hairpin tip (*shown on the left*), creates a single-stranded overhang that is complementary to the terminal coding sequence. Incorporation of this overhang into the coding junction produces a P nucleotide insert. If the single-stranded overhang is removed (not shown) or if the nick occurs directly at the hairpin tip (*shown on the right*), no P nucleotides are formed

stranded tail that may be incorporated into the junction to generate a P nucleotide insert. Since P nucleotides are only found adjacent to intact (untrimmed) coding ends (Lafaille et al. 1989; Meier and Lewis 1993), the covalently sealed coding ends should retain the entire coding sequence. This prediction has been verified, as the hairpins formed in scid thymocytes at the TCR Dδ2 segment contain the entire sequence of the Dδ2 coding element, without loss or addition of nucleotides (Zhu and Roth 1995). In support of the hairpin model, insertions resembling P nucleotides can be derived from opening of artificially constructed hairpins introduced into cultured cells by transfection (Lewis 1994b).

Since P nucleotides are not present at all coding joints, it is not clear whether hairpins are obligatory intermediates that do not always give rise to P nucleotides or whether there might be another pathway for forming coding joints that does not involve hairpins. The presence of P nucleotides in as many as 80% of junctions with undeleted coding ends (Meier and Lewis 1993) suggests that hairpin coding ends can be frequent intermediates in normal cells. However, the frequency of P nucleotide insertions at different coding segments varies considerably (Meier and Lewis 1993). Coding joints without P nucleotide inserts could be derived from hairpin coding ends through opening of the hairpin precisely at the tip (between the two terminal nucleotides), leaving a blunt end that cannot give rise to P nucleotides. Alternatively, potential P nucleotides could be lost by removal of the single-stranded extension prior to joining. Loss of single-stranded extensions is a common feature of end-joining in mammalian cells (Roth and Wilson 1986, 1988). Recent work has shown that "single-strand specific" nucleases such as mung bean nuclease and P1 nuclease can open hairpins both at the tip (leaving blunt ends) and away from the tip (leaving short single-stranded extensions) (KABOTYANSKI et al. 1995). The site of hairpin opening by these enzymes is strongly influenced by the nucleotide sequence of the hairpin (KABOTYANSKI et al. 1995). Thus, both mechanisms might contribute to formation of coding joints that do not contain P nucleotides.

The evidence summarized in the preceding paragraphs strongly suggests that hairpin coding ends are frequent, if not obligatory, intermediates in V(D)J recombination. These hairpins must presumably be opened prior to joining, and their accumulation in scid cells suggests that the *scid* defect somehow interferes with the opening reaction. Hairpin opening activities and the possible role of the *scid* factor will be discussed in the following sections.

5 Putting It All Together: Recognition, Cleavage, and Joining

Conceptually, the V(D)J recombination reaction can be divided into three steps: recognition, cleavage, and joining. Factors capable of recognizing the RSS and catalyzing double-strand cleavage should be required for the first two steps. In the joining phase of the reaction, factors must recognize and repair the broken DNA ends to generate coding and signal joints. A great deal of progress has been made recently in the identification of protein components of the V(D)J recombination machinery, and progress in this rapidly moving area will be summarized in this section (see other chapters in this volume for more detailed information). Finally, we shall attempt to incorporate this information into a working model.

In general, recombination systems that recognize specific sites (site-specific recombination and transposition reactions) require the assembly of stable DNA-protein complexes involving recombinases bound to their recognition sequences. These highly regulated reactions make use of architecturally defined DNA-protein complexes involving interactions between pairs of recognition sequences to control the reaction pathway (MIZUUCHI 1992). This form of regulation may also apply to V(D)J recombination to ensure that no cleavage occurs at one RSS without the presence of a functional partner RSS, a situation that could have disastrous consequences. Several lines of evidence provide indirect support for such a mechanism. Studies of integrated artificial recombination substrates have suggested that initiation of recombination may require a pair of RSS (HENDRICKSON et al. 1991a), and experiments using extrachromosomal substrates have provided evidence that physical interactions between a pair of RSS (synapsis) are required for efficient recombination (SHEEHAN and LIEBER 1993). A requirement for synapsis prior to formation of double-strand breaks is also supported by analysis of broken DNA molecules in mouse thymocytes, in which cleavage at pairs of RSS but not at individual RSS, was detected (ROTH et al. 1992b). Several steps might be regulated by synapsis. For example, an interaction between the two RSS might be required for initiation of cleavage at either signal; alternatively, nicks or double-strand breaks might occur at one signal, with synapsis required for cleavage at the partner RSS. Such requirements for signal-signal interactions prior to cleavage might form the basis of the 12/23 rule.

One scheme for cleavage following recognition and synapsis of RSS is presented in Fig. 5. The analysis of V(D)J recombination-associated double-strand breaks, described in the preceding sections, indicates that cleavage is a conservative reaction, producing a hairpin coding end and a blunt signal end that both contain the entire sequence of the coding and RSS elements (ZHU and ROTH 1995). These results are consistent with a coupled cleavage mechanism in which hairpins are formed directly by the cleavage reaction that liberates signal ends (ZHU and ROTH 1995) (Fig. 6). Since coding joints

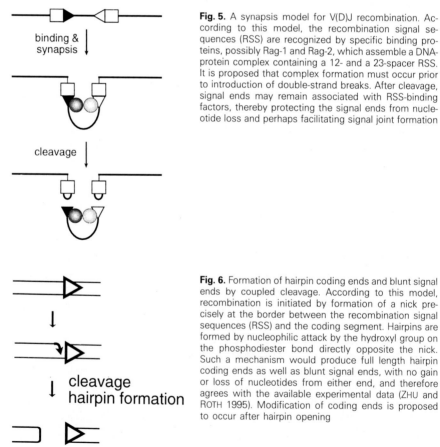

binding & synapsis

cleavage

Fig. 5. A synapsis model for V(D)J recombination. According to this model, the recombination signal sequences (RSS) are recognized by specific binding proteins, possibly Rag-1 and Rag-2, which assemble a DNA-protein complex containing a 12- and a 23-spacer RSS. It is proposed that complex formation must occur prior to introduction of double-strand breaks. After cleavage, signal ends may remain associated with RSS-binding factors, thereby protecting the signal ends from nucleotide loss and perhaps facilitating signal joint formation

cleavage
hairpin formation

Fig. 6. Formation of hairpin coding ends and blunt signal ends by coupled cleavage. According to this model, recombination is initiated by formation of a nick precisely at the border between the recombination signal sequences (RSS) and the coding segment. Hairpins are formed by nucleophilic attack by the hydroxyl group on the phosphodiester bond directly opposite the nick. Such a mechanism would produce full length hairpin coding ends as well as blunt signal ends, with no gain or loss of nucleotides from either end, and therefore agrees with the available experimental data (ZHU and ROTH 1995). Modification of coding ends is proposed to occur after hairpin opening

derived from these ends exhibit considerable nucleotide loss and addition, modification of coding ends probably occurs after hairpin opening.

Very recent experiments in which recognition and cleavage have been achieved in a cell-free system suggest that recognition involves the recombination activating genes, *Rag-1* and *Rag-2* (van GENT et al. 1995; see D.C. van GENT et al. this volume). A role of the Rag-1 and Rag-2 proteins in the formation of double-strand breaks is also supported by in vivo studies in Rag-deficient (knockout) mice, which neither signal ends nor coding ends can be detected (SCHLISSEL et al. 1993). However, since cleavage adjacent to an RSS in a cell-free system (a reaction that depends upon Rag-1 and Rag-2) does not require the presence of a second RSS (van GENT et al. 1995), additional factors may also participate in regulation of cleavage, for example through enforcement of the 12/23 rule.

Once cleavage has occurred, signal and coding ends must be joined. As discussed previously, one obvious difference between the handling of signal and coding ends is the precise nature of signal joint formation. The precision with which signal ends are handled contrasts with the much lower fidelity of end joining seen in transfected or microinjected DNA molecules. For example, in one study only 64% of junctions resulting from circularization of a blunt-ended substrate microinjected into the nuclei of fibroblastoid cells did not suffer modification of at least one end (ROTH et al. 1991). The much higher precision of signal end joining suggests that these ends may be specifically protected from nucleotide loss, presumably by noncovalent DNA-protein interactions that could involve Rag-1 and/or Rag-2 or perhaps other RSS binding factors. It is notable that addition of N nucleotides at signal joints, which is presumably mediated by TdT, occurs much more frequently than nucleotide loss (LIEBER et al. 1988b). Thus, whatever mechanism protects signal ends from nucleotide loss does not exclude TdT.

Another fundamental difference between the joining of signal and coding ends is the presumed requirement for a hairpin opening step prior to further processing of coding ends. Hairpin opening activity could be provided by a specific enzyme used in V(D)J recombination or by "nonspecific" activities capable of cleaving single-stranded DNA. Once hairpins are opened, the ends are available for modification, and ultimately are joined to produce a wide variety of junction structures, possibly employing mechanisms similar to those that mediate random end joining (ROTH and WILSON 1988).

The recent identification of several protein factors that presumably participate in end processing and joining was made possible by the isolation of mutations that affect V(D)J recombination. As previously mentioned, the *scid* mutation confers defects in both general DNA repair and V(D)J recombination. Although coding joint formation is severely impaired, the mutation has much less effect on signal joint formation (LIEBER et al. 1988a). Additional mutant cell lines with defects in both double-strand break repair and V(D)J recombination include XR-1 (XRCC4 complementation group; see Z. LI and F.W.ALT, this volume), xrs-5 and xrs-6 (XRCC5 complementation group; see G. CHU, this volume), and sxi-1 (PERGOLA et al. 1993; TACCIOLI et al. 1993; LEE et al. 1995; see S.E. LEE et al., this volume). Unlike the *scid* defect, these mutations impair both signal joint and coding joint formation, suggesting that they may be involved in end joining.

Recent studies (KIRCHGESSNER et al. 1995; BLUNT et al. 1995; see P.A. JEGGO et al. this volume) have indicated that the *scid* gene (XRCC7) encodes the catalytic subunit of the DNA-dependent protein kinase (DNA-PK$_{cs}$), a kinase which is activated in the presence of DNA ends. Remarkably, it appears that two of remaining three mutations known to impair both V(D)J recombination and double-strand break repair affect the DNA binding subunits of DNA-PK. Thus, XRCC5 encodes Ku80 (BOUBNOV et al. 1995; TACCIOLI et al. 1994; SMIDER et al. 1994), and some evidence suggests that the sxi-1 mutation may affect Ku70 (WEAVER 1995). The catalytic subunit of DNA-PK is activated by binding to a DNA-bound Ku heterodimer (DVIR et al. 1992; GOTTLIEB and JACKSON 1993; MOROZOV et al. 1994). In vitro studies have established that Ku is capable of binding both to double-strand breaks and to hairpins (MIMORI and HARDIN 1986; PAILLARD and STRAUSS 1991), suggesting that both coding and signal ends could serve to activate DNA-PK. Possible roles of these factors in V(D)J recombination will be discussed below.

6 How Does the *scid* Defect Block Formation of Coding Joints?

The specific accumulation of covalently sealed coding ends in scid thymocytes suggests that the *scid* defect somehow interferes with the hairpin opening reaction. The possibility that scid cells are generally incapable of opening hairpins has been invalidated by the observation that artificial hairpins introduced into scid cell lines by transfection are opened normally (LEWIS 1994b; STAUNTON and WEAVER 1994). However, this result does not preclude the possibility that the *scid* mutation specifically affects the opening of covalently sealed coding ends. First, it is not known whether the nuclease(s) responsible for opening transfected hairpins are also involved in the metabolism of coding ends. In fact, the inserts resembling P nucleotides that are derived from transfected hairpins tend to be longer than P nucleotides created from authentic coding ends (LEWIS 1994b), suggesting that two different nucleases might be responsible for processing these two types of hairpins. Second, hairpin coding ends may be assembled into specific DNA-protein complexes during V(D)J recombination. Perhaps the *scid* defect results in sequestration of the sealed coding ends in a DNA-protein complex that is inaccessible to the hairpin opening machinery (ROTH et al. 1995; ZHU and ROTH 1995). According to this proposal, the *scid* factor promotes the accessibility of hairpin coding ends, either by altering the complex or by "recruiting" a specific hairpin opening nuclease into the complex.

Now that the *scid* gene has been identified, it is possible to pose the accessibility model in more molecular (although no less hypothetical) terms. As shown in Fig. 7, we speculate that, after cleavage at the RSS, the coding and signal ends are both present in a DNA-protein complex that includes Ku

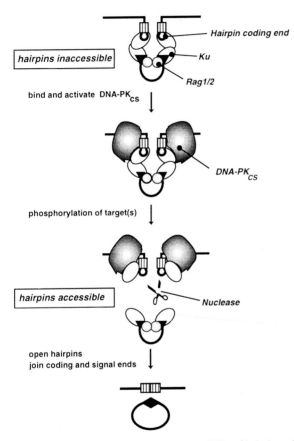

Fig. 7. Does the *scid* factor modulate accessibility of hairpin coding ends? According to this model, double-strand breaks are formed in the context of a synaptic complex involving recombination factors (possibly including Rag-1 and Rag-2) and a pair of signals. Cleavage generates a pair of signal ends (which may remain associated with recombination signal sequence-binding proteins) and a pair of hairpin coding ends. As shown in the figure, Ku binds to both the signal ends and the hairpin coding ends, forming complexes that prevent hairpin opening factors from gaining access to coding ends. The catalytic subunit of DNA-dependent protein kinase (DNA-PKcs) interacts with Ku, resulting in its activation. DNA-PKcs then phosphorylates target molecules, resulting in increased accessibility of coding ends to hairpin opening factors. After opening, coding ends are processed and joined to form coding joints. According to this model, in the absence of functional DNA-PKcs, such as in scid mice, coding ends are arrested in the hairpin stage, although signal ends can be joined by a pathway that is not completely dependent upon DNA-PKcs activity. See text for further discussion

and perhaps other components. The DNA-Ku complex then recruits the catalytic subunit which becomes activated upon binding. The active kinase could then modify target molecules, presumably by phosphorylation. Potential targets include components of the complex, such as Ku and DNA-PK$_{cs}$ itself, as well as other molecules. Modification of complex components could result in increased accessibility of the hairpins to the nuclease activity. Alternatively, the kinase could act on components outside the complex to recruit a specific hairpin-opening nuclease into the complex. In scid cells, the absence of DNA-PK$_{cs}$ activity results in failure of the coding ends to become accessible for opening, resulting in their accumulation. One attractive feature of this model is that it accounts for the accumulation of hairpin coding ends in cells that appear to be capable of efficiently opening exogenously introduced hairpins. The model is also consistent with another peculiar feature that has been observed at some of the rare coding joints isolated from scid cells: the presence of very long (up to 15 nucleotide) P nucleotide inserts (KIENKER et al. 1991; SCHULER et al. 1991; KOTLOFF et al. 1993). This is precisely what would be expected if the hairpin termini are protected by bound protein, so that opening is accomplished by nicking further from the hairpin tip.

How does this proposal address the defects in double-strand break repair conferred by mutations in Ku70, Ku80, and DNA-PK$_{cs}$? It is not difficult to imagine that the DNA-dependent protein kinase may play a role similar to those proposed above for V(D)J recombination. For example, DNA-PK might function in organizing DNA-protein complexes involved in joining broken ends or in signaling the recruitment of additional factors necessary for repair of particular lesions. Further work in this rapidly moving area will undoubtedly reveal more precisely the functions of these proteins and is likely to uncover additional factors involved in V(D)J recombination and double-strand break repair.

Acknowledgments. We thank Mary Lowe for excellent secretarial assistance. Research in the author's laboratory is supported by grants from the NIH (AI-36420) and the American Cancer Society (DB-118). D.B.R. is a Charles E. Culpeper Medical Scholar.

Note Added in Proof. Recent studies both in cell-free systems (EASTMAN QM, LEU TMJ and SCHATZ D. Nature 380: 85–88; VAN GENT DC, RAMSDEN DA and GELLERT M. Cell 85: 107–113) and in vivo (STEEN SB, GOMELSKY L and ROTH DB. Genes to Cells, in press) have demonstrated that efficient cleavage requires a 12/23 RSS pair. These data support a synapsis model and show that the 12/23 rule is enforced at or prior to the cleavage step.

References

Alt FW, Baltimore D (1982) Joining of immunoglobulin heavy chain gene segments: implications from a chromosome with evidence of three D-JH fusions. Proc Natl Acad Sci USA 79:4118–4122
Bainton R, Gamas P, Craig NL (1991) Tn7 transposition in vitro proceeds through an excised transposon intermediate generated by staggered breaks in DNA. Cell 65:805–816

Biedermann KA, Sun J, Giaccia AJ, Tosto LM, Brown JM (1991) scid mutation in mice confers hypersensitivity to ionizing radiation and a deficiency in DNA double-strand break repair. Proc Natl Acad Sci USA 88:1394–1397

Blunt T, Finnie NJ, Taccioli GE, Smith GCM, Demengeot J, Gottlieb TM, Mizuta R, Varghese AJ, Alt FW, Jeggo PA, Jackson SP (1995) Defective DNA-dependent protein kinase activity is linked to V(D)J recombination and DNA repair defects associated with the murine scid mutation. Cell 80:813–823

Bosma GC, Custer RP, Bosma MJ (1983) A severe combined immunodeficiency mutation in the mouse. Nature 301:527–530

Boubnov NV, Hall KT, Wills Z, Lee SE, He DM, Benjamin DM, Pulaski CR, Band H, Reeves W, Hendrickson EA, Weaver DT (1995) Complementation of the ionizing radiation sensitivity DNA end binding and V(D)J recombination defects of double-strand break repair mutants by the p86 Ku autoantigen. Proc Natl Acad Sci USA 92:890–894

Cao L, Alani E, Kleckner N (1990) A pathway for generation and processing of double-strand breaks during meiotic recombination in S. cerevisiae. Cell 61:1089–1101

Chang C, Biedermann KA, Mezzina M, Brown JM (1993) Characterization of the DNA double strand break repair defect in scid mice. Cancer Res 53:1244–1248

Connolly B, White CI, Haber JE (1988) Physical monitoring of mating type switching in Saccharomyces cerevisiae. Mol Cell Biol 8:2342–2349

Dvir A, Peterson SR, Knuth MW, Lu H, Dynan WS (1992) Ku autoantigen is the regulatory component of a template-associated protein kinase that phosphorylates RNA polymerase II. Proc Natl Acad Sci USA 89:11920–11924

Fulop GM, Phillips RA (1990) The scid mutation in mice causes a general defect in DNA repair. Nature 347:479–482

Gilfillan S, Dierich A, Lemeur M, Benoist, C Mathis D (1993) Mice lacking TdT: mature animals with an immature lymphocyte repertoire. Science 261:1175–1178

Gottlieb TM, Jackson SP (1993) The DNA-dependent protein kinase: requirement for DNA ends and association with Ku antigen. Cell 72:131–142

Haber JE (1992) Exploring the pathways of homologous recombination. Curr Opin Cell Biol 4:401–412

Haniford DB, Benjamin HW, Kleckner N (1991) Kinetic and structural analysis of a cleaved donor intermediate and a strand transfer intermediate in Tn10 transposition. Cell 64:171–179

Harrington J, Hsieh C-L, Gerton J, Bosma G, Lieber MR (1992) Analysis of the defect in DNA end joining in the murine scid mutation. Mol Cell Biol 12:4758–4768

Hendrickson EA, Schatz DG, Weaver DT (1988) The scid gene encodes a *trans*-acting factor that mediates the rejoining event of Ig gene rearrangement. Genes Dev 2:817–829

Hendrickson EA, Liu VF, Weaver DT (1991a) Strand breaks without DNA rearrangement in V(D)J recombination. Mol Cell Biol 11:3155–3162

Hendrickson EA, Qin X-Q, Bump EA, Schatz DG, Oettinger M, Weaver DT (1991b) A link between double-strand break-related repair and V(D)J recombination: the scid mutation. Proc Natl Acad Sci USA 88:4061–4065

Kabotyanski EB, Zhu C, Roth DB (1995) Hairpin opening by single-strand specific nucleases. Nucleic Acids Res 23:3872–3881

Kienker LJ, Kuziel WA, Tucker PW (1991) T cell receptor γ and δ gene junctional sequences in scid mice: excessive P nucleotide insertion. J Exp Med 174:769–773

Kirchgessner CU, Patil CK, Evans JW, Cuomo CA, Fried LM, Carter T, Oettinger MA, Brown JM (1995) DNA-dependent kinase (p350) as a candidate gene for the murine scid defect. Science 267:1178–1183

Komori T, Okada A, Stewart V, Alt FW (1993) Lack of N regions in antigen receptor variable region genes of TdT-deficient lymphocytes. Science 261:1171–1175

Kotloff DB, Bosma MJ, Ruetsch NR (1993) V(D)J recombination in peritoneal B cells of leaky scid mice. J Exp Med 178:1981–1994

Lafaille JJ, DeCloux A, Bonneville M, Takagaki Y, Tonegawa S (1989) Junctional sequences of T cell receptor γδ genes: implications for γδT cell lineages and for a novel intermediate of V(D)J joining. Cell 59:859–870

Lee SE, Pulaski CR, He DM, Benjamin DJ, Voss M, Um J, Hendrickson EA (1995) Isolation of mammalian cell mutants that are X-ray sensitive impaired in DNA double-strand break repair and defective for V(D)J recombination. Mutat Res 336:279–291

Lewis SM (1994a) P nucleotide insertions and the resolution of hairpin DNA structures in mammalian cells. Proc Natl Acad Sci USA 91:1332–1336

Lewis SM (1994b) The mechanism of V(D)J joining: lessons from molecular immunological and comparative analyses. Adv Immunol 56:27–150

Lieber MR, Hesse JE, Lewis S, Bosma GC, Rosenberg N, Mizuuchi K, Bosma MJ, Gellert M (1988a) The defect in murine severe combined immune deficiency: joining of signal sequences but not coding segments in V(D)J recombination. Cell 55:7–16

Lieber MR, Hesse JE, Mizuuchi K, Gellert M (1988b) Lymphoid V(D)J recombination: nucleotide insertion at signal joints as well as coding joints. Proc Natl Acad Sci USA 85:8588–8592

Malynn BA, Blackwell TK, Fulop GM, Rathbun GA, Furley AJW, Ferrier P, Heinke LB, Phillips RA, Yancopoulos GD, Alt FW (1988) The scid defect affects the final step of the immunoglobulin VDJ recombinase mechanism. Cell 54:453–460

McCormack WT, Tjoelker LW, Carlson LM, Petryniak B, Barth CF, Humphries EH, Thompson CB (1989) Chicken IgL gene rearrangement involves deletion of a circular episome and addition of single nonrandom nucleotides to both coding segments. Cell 56:785–791

Meier JT, Lewis SM (1993) P nucleotides in V(D)J recombination: a fine-structure analysis. Mol Cell Biol 13:1078–1092

Mimori T, Hardin JA (1986) Mechanism of Interaction between Ku Protein and DNA. J. Biol Chem 261:10375–10379

Mizuuchi K (1992) Transpositional recombination: mechanistic insights from studies of Mu and other elements. Annu Rev Biochem 61:1011–1051

Morozov VE, Falzon M, Anderson CW, Kuff EL (1994) DNA-dependent protein kinase is activated by nicks and larger single-stranded gaps. J Biol Chem 269:16684–16688

Mueller PR, Wold B (1989) In vivo footprinting of a muscle specific enhancer by ligation mediated PCR. Science 246:780–786

Paillard S, Strauss F (1991) Analysis of the mechanism of interaction of simian Ku protein with DNA Nucleic Acids Res 19:5619–5624

Pergola F, Zdzienicka MZ, Lieber MR (1993) V(D)J recombination in mammalian mutants defective in DNA double-strand break repair. Mol Cell Biol 13:3464–3471

Ramsden DA, Gellert M (1995) Formation and resolution of double strand break intermediates in V(D)J rearrangement. Genes Dev 9:2409–2420

Raveh D, Hughes SH, Shafer BK, Strathern JN (1989) Analysis of the HO-cleaved MAT DNA intermediate generated during the mating type switch in the yeast Saccharomyces cerevisiae. Mol Cell Biol 220:33–42

Roth DB, Proctor GN, Stewart LK, Wilson JH (1991) Oligonucleotide capture during end joining in mammalian cells. Nucleic Acids Res 19:7201–7205

Roth DB, Menetski JP, Nakajima PB, Bosma MJ, Gellert M (1992a) V(D)J recombination: broken DNA molecules with covalently sealed (hairpin) coding ends in scid mouse thymocytes. Cell 70:983–991

Roth DB, Nakajima PB, Menetski JP, Bosma MJ, Gellert M (1992b) V(D)J recombination in mouse thymocytes: double-strand breaks near T cell receptor δ rearrangement signals. Cell 69:41–53

Roth DB, Zhu C, Gellert M (1993) Characterization of broken DNA molecules associated with V(D)J recombination. Proc Natl Acad Sci USA 90:10788–10792

Roth DB, Wilson JH (1986) Nonhomologous recombination in mammalian cells: role for short sequence homologies in the joining reaction. Mol Cell Biol 6:4295–4304

Roth DB, Wilson JH (1988) Illegitimate recombination in mammalian cells. In: Genetic Recombination R Kucherlapati and G Smith (eds) ASM Press Washington DC pp 621–653

Roth DB, Lindahl T, Gellert M (1995) How to make ends meet. Curr Biol 5:496–499

Ruan K-S, Emmons SW (1994) Extrachromosomal copies of transposon Tc1 in the nematode Caenorhabditis elegans. Proc Natl Acad Sci USA 81:4018–4022

Schlissel M, Constantinescu A, Morrow T, Baxter M, Peng A (1993) Double-strand signal sequence breaks in V(D)J recombination are blunt 5′-phosphorylated RAG-dependent and cell cycle regulated. Genes Dev 7:2520–2532

Schuler W, Weiler IJ, Schuler A, Phillips RA, Rosenberg N, Mak TW, Kearney JF, Perry RP, Bosma MJ (1986) Rearrangement of antigen receptor genes is defective in mice with severe combined immune deficiency. Cell 46:963–972

Schuler W, Ruetsch NR, Amsler M, Bosma MJ (1991) Coding joint formation of endogenous T cell receptor genes in lymphoid cells from scid mice: unusual P nucleotide additions in VJ-coding joints. Eur J Immunol 21:589–596

Sheehan KM, Lieber MR (1993) V(D)J recombination: signal and coding joint resolution are uncoupled and depend on parallel synapsis of the sites. Mol Cell Biol 13:1363–1370

Smider V, Rathmell WK, Lieber MR, Chu G (1994) Restoration of X-ray resistance and V(D)J recombination in mutant cells by Ku cDNA. Science 266:288–291

Staunton JE, Weaver DT (1994) scid cells efficiently integrate hairpin and linear DNA substrates. Mol Cell Biol 14:3876–3883

Sun H, Treco D, Schultes NP, Szostak JW (1989) Double-strand breaks at an initiation site for meiotic gene conversion. Nature 338:87–90

Taccioli GE, Rathbun G, Oltz E, Stamato T, Jeggo PA, Alt FW (1993) Impairment of V(D)J recombination in double-strand break repair mutants. Science 260:207–210

Taccioli GE, Gottlieb TM, Blunt T, Priestley A, Demengeot J, Mizuta R, Lehmann AR, Alt FW, Jackson SP, Jeggo PA (1994) Ku80: product of the XRCC5 gene and its role in DNA repair and V(D)J recombination. Science 265:1442–1445

van Gent DC, McBlane JF, Ramsden DA, Sadofsky MJ, Hesse JE, Gellert M (1995) Initiation of V(D)J recombination in a cell-free system Cell 81:925–934

van Luenen HGAM, Colloms SD, Plasterk RHA (1993) Mobilization of quiet endogenous Tc3 transposons of Caenorhabditis elegans by forced expression of Tc3 transposase. EMBO J 12:2513–2520

Weaver DT (1995) V(D)J recombination and double-strand break repair. Adv Immunol 58:29–85

White CI, Haber JE (1990) Intermediates of recombination during mating type switching in Saccharomyces cerevisiae. EMBO J 9:663–673

Zhu C, Roth DB (1995) Characterization of coding ends in thymocytes of scid mice: implications for the mechanism of V(D)J recombination. Immunity 2:101–112

Identification of the Catalytic Subunit of DNA Dependent Protein Kinase as the Product of the Mouse *scid* Gene

Penny A. Jeggo[1], Stephen P. Jackson[2] and Guillermo E. Taccioli[3]

1 Introduction

A severe combined immunodeficient (*scid*) mouse was identified by Bosma in 1983 (Bosma et al. 1983) and found to lack mature T and B cells. Subsequently, the immune deficiency of *scid* mice was determined to be due to an inability to undergo effective V(D)J recombination, a process which occurs during T and B cell development and effects the reassortment of the variable (V), diversity (D) and joining (J) segments into a contiguous and functional exon (Schuler et al. 1986; Lieber et al. 1988; Malynn et al. 1988; Blackwell et al. 1989; Bosma and Carroll 1991). In germ line cells, each V, D and J segment is associated with a partially conserved recombination signal sequence (RSS), and the first step in V(D)J recombination involves the introduction of two double strand breaks (dsbs) between two RSS and their associated V, D or J coding sequences (for reviews, see Alt et al. 1992; Gellert 1992; Lewis 1994). The ends of these two dsbs are recombined to yield an RSS junction, which represents the

[1] MRC Cell Mutation Unit, University of Sussex, Brighton, BN1 9RR, UK
[2] Wellcome/CRC Institute and Department of Zoology, Cambridge University, Tennis Court Road, Cambridge, CB2 1QR, UK
[3] Howard Hughes Medical Institute, The Childrens Hospital and Department of Genetics and Center for Blood Research, Harvard University Medical School, Boston, MA 02115, USA

precise joining of two RSS sequences, and a coding junction, at which two coding sequences are rejoined. In contrast to the RSS junctions, however, coding joint formation is invariably imprecise with small nucleotide deletions or additions at the junctions, which can arise by a number of different mechanisms. This imprecision in coding joint formation, coupled with the reassortment of the V, D, or J segments, provide mechanisms responsible for the enormous diversity of the immune response.

Scid mice have an interesting phenotype, in that RSS join formation occurs almost normally whilst the frequency of coding joins is enormously impaired. Since coding joint formation is the step which produces a functional exon, a defect in this stage of the reaction will result in a non-functional T or B cell. Two other manifestations of the scid phenotype are the accumulation of coding ends as hairpin structures (ROTH et al. 1992; ZHU and ROTH 1995) and an enhanced length of so-called palindromic (P) nucleotides at the coding junctions that do form (LAFAILLE et al. 1989; KIENKER et al. 1991; SCHULER et al. 1991). P nucleotides are one form of nucleotide insertions seen in coding joins of normal cells, which can be identified as the inverse complement of the terminal junction sequences. Since hairpin structures provide a convenient explanation for the formation of p nucleotides, it has been argued that they represent an intermediate structure in the recombination process in normal cells, even though such structures have not hitherto been identified in normal cells (LIEBER 1991; ZHU and ROTH 1995). In exciting recent work, van Gent and co-workers have established conditions in which cell-free extracts will carry out the initial cleavage step of V(D)J recombination, and have found hairpin structures at the termini of coding sequences, providing support that such structures do arise during or directly following the initial cleavage reaction (VAN GENT et al. 1995).

In lower organisms a strong correlation exists between recombination and dsb repair; most pathways of recombination utilise a dsb as an intermediate, and most recombination defective mutants are defective in their ability to rejoin DNA dsbs, a correlation exemplified by rad52 mutants of yeast (JEGGO 1990; FRIEDBERG 1991). These links prompted an examination of scid cells for their response to ionising radiation (IR), a DNA damaging agent which induces the formation of DNA dsbs. Scid cells proved to be exquisitely radiosensitive, and defective in DNA dsb rejoining (FULOP and PHILLIPS 1990; BIEDERMANN et al. 1991; HENDRICKSON et al. 1991). The correlation between inability to rejoin DNA dsbs and defective V(D)J recombination was also established from the converse direction. A number of radiosensitive mutants derived from established rodent cell lines had been characterised as having defects in their ability to rejoin DNA dsbs (KEMP et al. 1984; GIACCIA et al. 1985; WHITMORE et al. 1989). An examination of these mutants for their ability to carry out V(D)J recombination showed the same correlation, whilst, in contrast, other radiosensitive mutants proficient in dsb rejoining were likewise proficient in V(D)J recombination (PERGOLA et al. 1993; TACCIOLI et al. 1993, 1994a). Fusion analysis between these mutants has demonstrated the existence of four complementation groups, more recently referred to as groups 4–7 (JEGGO et al. 1991; THACKER and WILKINSON 1991; THOMPSON and JEGGO 1995). Signficantly, one of the hamster mutants (V-3) isolated by its extreme radiosensitivity, is defective in the same complementation group as *scid* cells (TACCIOLI et al. 1994a). Despite showing the same overall phenotype, the precise details of the defects differ for each complementation group. Most noticeably, mutants belonging to groups 4–6 harbour defects in the frequency and accuracy of both signal and coding joint formation, and this contrasts markedly with both of the group 7 mutants (*scid* and V-3) which exhibit only a minor defect in signal join formation (see above) (PERGOLA et al. 1993; TACCIOLI et al. 1994a).

2 Chromosomal Localisation of the *SCID (XRCC7)* Gene

The marked radiosensitivity of *scid* cells provided a route to select for complemented cells. Attempts to clone the *SCID* gene by DNA transfection therefore ensued, but with little success. Positional cloning strategies were therefore adopted by which a complementing human gene was localised to the pericentric region of the q arm of chromosome 8 (ITOH et al. 1993; KIRCHGESSNER et al. 1993). Complementation by chromosome 8 has been observed for both the radiosensitivity and V(D)J recombination defect of *scid* cells (BANGA et al. 1994).

3 Identification of Ku80 as the Product of *XRCC5*

The genes defective in complementation groups 4–7 have been designated *XRCC4–7*. Attempts to clone *XRCC4*, 5 and 6 were undertaken in parallel to the above studies. Human genes complementing groups 4 and 5 mutants were shown to map to chromosome 5 and 2q34-36, respectively (GIACCIA et al. 1990; JEGGO et al. 1992; HAFEZPARAST et al. 1993). Group 5 mutants were also characterised as lacking a DNA end-binding activity, with properties resembling Ku, an abundant nuclear protein identified in 1986 as an autoantigen present in the sera of certain autoimmune patients (MIMORI et al. 1981, 1986; GETTS and STAMATO 1994; RATHMELL and CHU 1994a, 1994b). Ku is a heterodimer containing subunits of 70 and 80 kDa, and excitement rose when the 80 kDa subunit was also shown to map to the region 2q33-35 (CAI et al. 1994). The ability of Ku80 cDNA to complement the radiosensitivity and defects in dsb rejoining, V(D)J recombination and DNA end-binding activity of the IR group 5 mutants provided confirmation that Ku80 is the product of *XRCC5* (SMIDER et al. 1994; TACCIOLI et al. 1994b; BOUBNOV et al. 1995). Interestingly, none of these defects was fully regained, suggesting that the human homologue could not fully substitute for the hamster protein, a feature indicating that Ku80 may interact as part of a protein complex. Significantly, Ku has recently been shown to be the DNA end-binding component of DNA dependent protein kinase (DNA-PK). The catalytic component of DNA-PK (DNA-PK$_{cs}$) is a large polypeptide of approximately 450 kDa, which contains a serine-threonine kinase domain at its 3' terminus. As the name suggests, activation of the kinase function is dependent upon binding (via Ku) to ds DNA ends. The involvement of Ku80 in DNA repair and V(D)J recombination therefore incriminated the two additional DNA-PK component proteins, Ku70 and DNA-PK$_{cs}$, as potential factors defective in group 7 mutants.

4 Identification of DNA-PK$_{cs}$ as the Product of *XRCC7*

4.1 Biochemical Examination of Group 7 Mutants

V-3 and *scid* cells were also examined for Ku-associated DNA end-binding activity, and found to contain parental levels, suggesting that they were not defective in the Ku70 component of Ku (GETTS and STAMATO 1994; RATHMELL and CHU 1994a). The next step was therefore to examine DNA-PK activity in these cell lines. Ku is a highly abundant nuclear protein in human cells, but its levels are much reduced in rodent cell lines. Similarly, whilst DNA-PK activity in human cells can be easily assayed by measuring the difference between phosphorylation of a suitable peptide in the presence and absence of DNA, this assay in its original form was not sufficiently sensitive to detect the much smaller levels of DNA-PK activity present in rodent extracts. Indeed, it was actually unclear whether rodent cells possessed DNA-PK activity. FINNIE et al. (1995) overcame this problem by developing a more sensitive assay for DNA-PK, in which DNA-binding proteins are concentrated by a DNA-pull down step, and phosphorylation of a p53 derived peptide is used as an assay. This simple purification step reduces the background phosphorylation due to other cellular kinases, and allows the detection and quantification of DNA-PK activity in rodent cell extracts that was estimated to be approximately 50-fold less than that present in human cell extracts. PETERSON et al. (1995) employed a slightly different assay involving the phosphorylation of the murine C-terminal domain subunit of RNA polymerase II fused to the DNA-binding domain of the yeast GAL4 protein that appeared to be sufficiently sensitive to detect the lower levels of DNA-PK activity in rodent cell extracts. Significantly when both assays were employed with extracts from *scid* and V-3 cells, no DNA-PK activity could be detected (BLUNT et al. 1995; PETERSON et al. 1995). Similarly, DNA-PK activity was lacking in extracts of *xrs-6* cells, a result predicted by the defect in Ku80 in these cells (FINNIE et al. 1995). This result nevertheless confirmed in vitro studies suggesting that Ku is the major, and possibly only, DNA-binding component of DNA-PK.

The lack of DNA-PK activity in group 7 mutants, therefore, suggested the exciting possibility that they might be defective in DNA-PK$_{CS}$. Mixing of extracts from group 5 and 7 mutants resulted in biochemical complementation of DNA-PK activity, in contrast to the lack of complementation seen when extracts from the two group 7 mutants were mixed (BLUNT et al. 1995). These results correlated with the results of complementation carried out by cell fusion, and demonstrated that, whilst group 5 and 7 mutants both phenotypically lack DNA-PK activity, they carry distinct genetic defects.

4.2 Examination of Human/Mouse Hybrids Complementing the *scid* Defect

In studies aimed at cloning the *XRCC7* gene by a positional cloning strategy, KIRCHGESSNER et al. (1995) constructed a panel of 24 mouse/human hybrid clones containing fragments of human chromosome 8. Some (16) of these hybrids complemented the radiosensitivity defects of *scid* cells, others (8) were non-complementing. In this way the *XRCC7* gene was localised to a 2 Mb segment mapping to the centromeric region of the long arm of chromosome 8. In order to examine whether the human DNA-PK$_{CS}$ gene was present within this region, they examined expression of human DNA-PK$_{CS}$ by Western blot analysis in their panel of hybrid clones using anti-DNA-PK$_{CS}$ antibody (KIRCHGESSNER et al. 1995). They found complete correlation between expression of human DNA-PK$_{CS}$ protein and complementation of the radiosensitivity of *scid* cells. Additional confirmation that DNA-PK$_{CS}$ maps to this region of chromosome 8 was shown by Sipley et al (1995) using 'in situ' hybridisation of DNA-PK$_{CS}$ cDNA clones, and by BLUNT et al (1995) who showed that yeast artificial chromomes (YACs) encoding DNA-PK$_{CS}$ map to the centromeric region of chromosome 8 (BLUNT et al. 1995; SIPLEY et al. 1995).

4.3 Complementation of V-3 Cells by the Introduction of YACs Encoding the DNA-PK$_{CS}$ Gene

DNA-PK$_{CS}$is one of the largest cDNAs known (approximately 14 Kb), and a full length cDNA is not available. To examine whether the gene could complement the repair and V(D)J recombination defects of group 7 mutants, BLUNT et al. (1995) took an alternative route of introducing YACs encoding the DNA-PK$_{CS}$ gene into group 7 mutants. Complementation of the radiosensitive phenotype of V-3 cells was achieved with clones derived from two YACs expressing the DNA-PK$_{CS}$ gene. In all complementing clones obtained, the human DNA-PK$_{CS}$ gene appeared to have been transferred intact to the V-3 cells, as judged by Southern hybridisation using DNA-PK$_{CS}$ cDNA clones as probes, whilst deletions in the DNA-PK$_{CS}$ gene could be observed in the non-complementing clones. Only partial complementation was achieved for radiosensitivity but the same

clones were fully complemented for the defect in V(D)J recombination. Interestingly, these clones had restored DNA-PK activity to levels approaching that
present in human cells, and exceeding the level present in wild-type hamster
cells. DNA-PK$_{CS}$ protein of human origin was also expressed in the complementing clones. Clones non-complementing for survival did not regain kinase
activity, nor did they express the human protein.

4.4 Lack of Expression of Endogenous DNA-PK$_{CS}$ Protein in Group 7 Mutants

The analysis of DNA-PK$_{CS}$ protein levels by immunoblotting experiments have
been difficult to carry out with *scid* and V-3 cells, since DNA-PK$_{CS}$ is poorly
expressed in rodent cells, and/or since the available antibodies raised to the
human protein, might not cross-react strongly with the rodent protein. However,
using a variety of techniques to enhance the sensitivity, anti-DNA-PK$_{CS}$ signal
can be detected in extracts from wild-type mouse and hamster cell lines, but
significantly is not detectable in extracts from *scid* or V-3 cells, nor from primary
fibroblast cells derived from the *scid* mouse (KIRCHGESSNER et al. 1995; PETERSON
et al. 1995, and our unpublished observations). Additionally, indirect immunofluorescence and the more sensitive technique of enhanced chemilumines-
cence have also been used to examine DNA-PK$_{CS}$ expression and localisation
in *scid* and V-3 cell lines, and again no or greatly reduced levels of protein
were detectable (KIRCHGESSNER et al. 1995; PETERSON et al. 1995).

5 Nature of the Defect in *scid* Mice

These combined studies provide convincing evidence that DNA-PK$_{CS}$ can complement the defects in the two group 7 mutants (summarised in Table 1).
Taken together with the inability to detect DNA-PK activity and DNA-PK protein
in *scid* and V-3 cells, these data strongly suggest that these two cell lines
carry mutations in the DNA-PK$_{CS}$ gene. In addition, the gene for CEBPδ (CCAAT-
enhancing binding protein δ) maps to the pericentric region of human chromosome 8 and to mouse chromosome 16 with close linkage to the *lgl* locus,
thus establishing synteny between human chromosome 8 and mouse chromosome 16. From linkage studies with mice, the *SCID* gene also maps to
this region of chromosome 16. It should be noted, however, that verification
that the DNA-PK$_{CS}$ gene itself rather than a regulatory gene is defective still
requires the detection of a mutation within this large gene. All the studies
concord in showing that both *scid* and V-3 cells contain reduced levels of
material cross-reacting with DNA-PK$_{CS}$ antibodies. What is currently unclear,
however, is whether some residual level of DNA-PK$_{CS}$ protein remains, since

Table 1. Defects in *scid* and V-3 cells and complementation by YACs and chromosome 8

Cell Lines	IR sensitivity	DNA-PK activity	DNA-PK protein levels (by Western blotting)	V(D)J Recombination		
				Signal joins		Coding joins
7				Frequency	Accuracy	Frequency
3T3 or CB-17	R	+[a,b]	+[b,c]	1	1	1
scid (SCGR11)	S	not detectable[a,b]	low or not detectable[b,c]	0.1[c]	not reported	0.001[c]
scid + human 8	R[P]	+[,b]	++[c]/+[b]	0.5[c]	not reported	0.5[c]
AA8	R	+[a]	+[b]	1	1	1
V-3	S	not detectable[a]	low or not detectable[b]	0.22[a]	0.47[a]	0.013[a]
V-3 + yac	R[P]	+++[a]	+++[a]	2.2[a]	0.91[a]	2.0[a]

IR, ionising radiation; *R*, wild type level of resistance to IR; *S*, sensitive to IR; R^P, partial recovery of resistance to IR.
V(D)J recombination frequencies shown are expressed relative to the parental cell line (for specific details of the assay see the relevent publication). The plasmids employed in the studies shown by footnotes a and c differ, so that a comparison of relative frequency cannot be made between these data sets.
[a] BLUNT et al. (1995)
[b] PETERSON et al. (1995)
[c] KIRCHGESSNER et al. (1995)

faint signals can be observed in the immunoblot data presented by KIRCHGESSNER et al (1995). No residual DNA-PK activity is measurable in *scid* and V-3 extracts but insensitivity of the assay would only permit levels greater than one-tenth parental levels to be detected. Whether residual activity exists is significant to assessing the *scid* phenotype. *Scid* mice have been described as having a "leaky" phenotype since approximately one-fifth of mice undergo rearrangements producing functional T and B lymphocytes. Additionally, the analysis of coding joint formation, using the transient transfection assay on *scid* and V-3 cells demonstrates a less severe defect than that observed with group 4 and 5 mutants (PERGOLA et al. 1993; TACCIOLI et al. 1993, 1994a). It is perhaps significant that all of the group 5 mutants examined show a similar severe defect, whilst both V-3 and *scid* cells show phenotypically the same, slightly milder defect. It has also recently been reported that a single exposure of newborn *scid* mice to a low dose of IR, results in the restoration of V(D)J recombination in the T cell receptor β chain locus (TCRβ), although not in B cells (DANSKA et al. 1994). In this context, it is noteworthy that exposure of *scid* cells to low dose IR did not result in any induction of DNA-PK activity (our unpublished observations). These "leaky" characteristics could reflect a "leaky" mutation in the *SCID* (DNA-PK$_{cs}$) gene or a *scid*-independent route to gain functional coding rearrangements. The latter explanation could include the existence of an alternative pathway or could indicate a less-than-absolute requirement for the DNA-PK$_{cs}$ gene in the pathway characterised to date. It is also noteworthy, that the two group 7 mutants (*scid* and V-3) show a similar level of IR sensitivity to the group 4 and 5 mutants (*xrs* and XR-1), an explanation less supportive of a leaky mutation in the *SCID* gene. The data available to date therefore, do not allow these questions to be answered. The nature of the mutation in the DNA-PK$_{cs}$ gene in *scid* and V-3 cells will thus be of great interest, and will have to await sequence analysis of this large gene in the mutant and wildtype rodent cells.

6 Role of DNA-PK in DNA Repair and V(D)J Recombination

These data, in addition to identifying DNA-PK$_{cs}$ as the product of *XRCC7*, also establish a function for the well-characterised DNA-PK protein in DNA repair and V(D)J recombination. Suprisingly, however, the two types of junctions formed during V(D)J recombination have differing requirements for the component subunits of DNA-PK. RSS joint formation is dependent upon functional Ku protein but occurs in the absence of DNA-PK$_{cs}$. Coding joint formation requires both Ku and the large subunit. This demonstrates that Ku can function in the absence of DNA-PK$_{cs}$, and thus independently of its role in the DNA-PK complex. In addition, it shows that direct end-joining can occur in the absence of DNA-PK$_{cs}$, and distinguishes signal joint formation from coding joint formation and the rejoining of damage induced ends. One possible explanation is that at signal junctions a component(s) of the V(D)J recombination appparatus can carry out the functions performed by DNA-PK$_{cs}$. Another explanation is that DNA-PK$_{cs}$ is essential for the specific modifications of ends that occur at coding junctions and damage induced ends, but is not required for the simple ligation needed to rejoin signal ends. The ability of Ku to bind to DNA ends would suggest that it might function early in the repair/recombination process, and probably serves to protect the ends from nucleolytic degradation since the DNA-PK defective mutants have large deletions in the rare junctions that do form. The precise role of the kinase activity induced specifically following the introduction of DNA ends remains to be elucidated. The DNA-PK defective mutants arrest normally at the S and G2 cell cycle checkpoints following radiation (JEGGO 1985; WEIBEZAHN et al. 1985) suggesting that they are not defective in a unique signalling pathway affecting these checkpoint controls. Other possibilities for the role of the kinase activity include the phosphorylation of the repair/recombination machinery or metabolic processes, such as transcription, whose activity may hinder the repair process. However, DNA-PK$_{cs}$ may have a function independently of, or in addition to, its kinase activity. One possibility, compatible with the large size of this protein, is that

DNA-PK$_{CS}$ could play a structural role, acting as a framework to which other proteins involved in the repair/recombination pathway might associate.

It is intriguing that the levels of DNA-PK protein and kinase activity are 50-fold greater in human cells than in the rodent cells, yet the levels of radioresistance are similar. Moreover, V-3 cells complemented by the YACs encoding DNA-PK$_{CS}$ regain kinase activity to levels greater than that present in the wild-type rodent cells, and almost to the higher levels found in human cells (BLUNT et al. 1995). This shows firstly that the greater levels of DNA-PK$_{CS}$ found in human cells are regulated by sequences closely linked to this gene which can be recognised by rodent cells. Secondly, and surprisingly, despite having elevated kinase activity compared with wild-type mouse cells, complemented V-3 cells are only partially restored for radioresistance yet fully restored for their defect in V(D)J recombination. Similarly, *scid* cells complemented by chromosome 8 also regain partial radioresistance (KIRCHGESSNER et al. 1993, 1995). At present it is difficult to evaluate these differences but they are suggestive that DNA-PKcs might have a further role in DNA repair in addition to its kinase activity, and lend support to a role as a framework or structural protein.

In conclusion, these studies have identified *scid* and V-3 as lacking DNA-PK activity and DNA-PK$_{CS}$ protein. These features can be restored together with radioresistance and the ability to carry out V(D)J recombination by the introduction of sequences encoding the DNA-PK$_{CS}$ gene. These studies provide strong evidence that DNA-PK$_{CS}$ is the product of *XRCC7*, the gene defective in mouse *scid* cells.

Acknowledgements. We thank members of our laboratories who have contributed to this work including T. Blunt, A. Priestley, N. Finnie, T. Gottlieb, J. Demengeot, R. Mizuta, F.W. Alt and A. R. Lehmann. Research in the MRC CMU is supported by European Community grants F13PCT920007 and ERBSCICT920823 and in the SPJ laboratory by grants SP2143/0101 and SP2143/0201. PAJ and SPJ are also funded by a collaborature grant from Kay Kendall Leukemia Fund. GET is a special fellow of the Leukemia Society of America.

References

Alt FW, Oltz EM, Young F, Gorman J, Taccioli G, Chen J (1992) VDJ recombination. Immunol Today 13:306–314

Banga SS, Hall KT, Sandhu AK, Weaver DT, Athwal RS (1994) Complementation of V(D)J recombination defect and X-ray-sensitivity of *scid* mouse cells by human-chromosome-8. Mutat Res DNA Repair 315(3):239–247

Biedermann KA, Sun J, Giaccia AJ, Tosto LM, Brown JM (1991) *scid* mutation in mice confers hypersensitivity to ionizing radiation and a deficiency in DNA double-strand break repair. Proc Natl Acad Sci USA 88:1394–1397

Blackwell TK, Malynn BA, Pollock RR, Ferrier P, Covey LR, Fulop GM, Phillips RA, Yancopoulos GD, Alt FW (1989) Isolation of scid pre-B cells that rearrange kappa light chain genes: formation of normal signal and abnormal coding joints. EMBO J 8:735–742

Blunt T, Finnie NJ, Taccioli GE, Smith GCM, Demengeot J, Gottlieb TM, Mizuta R, Varghese AJ, Alt FW, Jeggo PA, Jackson SP (1995) Defective DNA-dependent protein kinase activity is linked to V(D)J recombination and DNA repair defects associated with the murine *scid* mutation. Cell 80:813–823

Bosma GC, Custer RP, Bosma MJ (1983) A severe combined immunodeficiency mutation in the mouse. Nature 301:527:

Bosma MJ, Carroll AM (1991) The SCID mouse mutant: definition, schcharacterization, and potential uses. Annu Rev Immunol 9:323–350

Boubnov NV, Hall KT, Wills Z, Sang EL, Dong MH, Benjamin DM, Pulaski CR, Band H, Reeves W, Hendrickson EA, Weaver DT (1995) Complementation of the ionizing radiation sensitivity, DNA end binding, and V(D)J recombination defects of double-strand break repair mutants by the p86 Ku autoantigen. Proc Natl Acad Sci USA 92(3):890–894

Cai Q-Q, Plet A, Imbert J, Lafage-Pochitaloff M, Cerdan C, Blanchard J-M (1994) Chromosomal location and expression of the genes coding for Ku p70 and p80 in human cell lines and normal tissues. Cytogenet Cell Genet 65:221–227

Danska JS, Pflumio F, Williams CJ, Huner O, Dick JE, Guidos CJ (1994) Rescue of T cell-specific V(D)J recombination in SCID mice by DNA-damaging agents. Science 226:450–455

Finnie NJ, Gottlieb TM, Blunt T, Jeggo PA, Jackson SP (1995) DNA-dependent protein-kinase activity is absent in xrs-6 cells implications for site-specific recombination and dna double-strand break repair. Proc Natl Acad Sci USA 92(1):320–324

Friedberg EC (1991) Yeast genes involved in DNA-repair processes: New looks on old faces. Mol Microbiol 5:2303–2310

Fulop GM, Phillips RA (1990) The scid mutation in mice causes a general defect in DNA repair. Nature 374:479–482

Gellert M (1992) Molecular analysis of V(D)J recombination. Annu Rev Genet 26:425–446

Getts RC, Stamato TD (1994) Absence of a Ku-like DNA end binding-activity in the xrs double-strand DNA repair-deficient mutant. J Biol Chem 269(23):15981–15984

Giaccia A, Weinstein R, Hu J, Stamato TD (1985) Cell cycle-dependent repair of double-strand DNA breaks in a gamma-ray sensitive Chinese hamster cell. Somatic Cell Mol Genet 11:485–491

Giaccia AJ, Denko N, MacLaren RA, Mirmen D, Waldren C, Hart I, Stamato TD (1990) Human chromosome 5 complements the DNA double-strand break-repair deficiency and gamma-ray sensitivity of the XR-1 hamster variant. Am J Hum Genet 47:459–469

Hafezparast M, Kaur GP, Zdzienicka M, Athwal RS, Lehmann AR, Jeggo PA (1993) Sub-chromosomal localisation of a gene (XRCC5) involved in double strand break repair to the region 2q34-36. Somatic Cell Mol Genet 19:413–421

Hendrickson EA, Qin X-Q, Bump EA, Schatz DG, Oettinger M, Weaver DT (1991) A link between double-strand break-related repair and V(D)J recombination: the scid mutation. Proc Natl Acad Sci USA 88:4061–4065

Itoh M, Hamatani K, Komatsu K, Araki R, Takayama K, Abe M (1993) Human chromosome-8 (p12-]q22) complements radiosensitivity in the severe combined immune-deficiency (scid) mouse. Radiat Res 134:364–368

Jeggo PA (1985) X-ray sensitive mutants of Chinese hamster ovary cell line: Radiosensitivity of DNA synthesis. Mutat Res 145:171–176

Jeggo PA (1990) Studies on mammalian mutants defective in rejoining double-strand breaks in DNA. Mutat Res 239:116

Jeggo PA, Tesmer J, Chen DJ (1991) Genetic analysis of ionising radiation sensitive mutants of cultured mammalian cell lines. Mutat Res 254:125–133

Jeggo PA, Hafezparast M, Thompson AF, Broughton BC, Kaur GP, Zdzienicka MZ, Athwal RS (1992) Localization of a DNA repair gene (XRCC5) involved in double-strand break rejoining to human chromosome 2. Proc Natl Acad Sci USA 89:6423–6427

Kemp LM, Sedgwick SG, Jeggo PA (1984) X-ray sensitive mutants of Chinese hamster ovary cells defective in double-strand break rejoining. Mutat Res 132:189–196

Kienker LJ, Kuziel WA, Tucker PW (1991) T cell receptor gamma and delta gene junctional sequences in SCID mice: Excessive P nucleotide insertion. J Exp Med 174(4):769–773

Kirchgessner CU, Tosto LM, Biedermann KA, Kovacs M, Araujo D, Stanbridge EJ, Brown JM (1993) Complementation of the radiosensitive phenotype in severe combined immunodeficient mice by human chromosome 8. Cancer Res 53(24):6011–6016

Kirchgessner CU, Patil CK, Evans JW, Cuomo CA, Fried LM, Carter T, Oettinger MA, Brown JM, Iliakis G, Mehta R, Jackson M (1995) DNA-dependent kinase (p350) as a candidate gene for murine SCID defect. Science 267(5201):1178–1183

Lafaille JJ, DeCloux A, Bonneville M, Takagaki Y, Tonegawa S (1989) Junctional sequences of T cell receptor gamma-delta genes: implications for gamma-delta T cell lineages and for a novel intermediate of V-(D)-J joining. Cell 59:859–870

Lewis SM (1994) The mechanism of V(D)J joining: Lessons from molecular, immunological, and comparative analyses. Adv Immunol 56:27–150

Lieber MR (1991) Site specific recombination in the immune system. FASEB J 5:29342944

Lieber MR, Hesse JE, Lewis S, Bosma GC, Rosenberg N, Mizuuchi K, Bosma MJ, Gellert M (1988) The defect in murine severe combined immune deficiency: joining of signal sequences but not coding segments in V(D)J recombination. Cell 55:7–16

Malynn BA, Blackwell TK, Fulop GM, Rathbun GA, Furley AJW, Ferrier P, Heinke LB, Phillips RA, Yancopoulos GD, Alt FW (1988) The scid defect affects the final step of the immunoglobulin VDJ recombinase mechanism. Cell 54(4):453–460

Mimori T, Akizuki M, Yamagata H, Inada S, Yoshida S, Homma M (1981) Characterization of a high molecular weight acidic nuclear protein recognized by autoantibodies in sera from patients with polymyositis-scleroderma overlap syndrome. J Clin Invest 68(3):611–620

Mimori T, Hardin JA, Steitz JA (1986) Characterization of the DNA-binding protein antigen Ku recognized by autoantibodies from patients with rheumatic disorders. J Biol Chem 261(5):2274–2278

Pergola F, Zdzienicka MZ, Lieber MR (1993) V(D)J recombination in mammalian cell mutants defective in DNA double strand break repair. Mol Cell Biol 13:3464–3471

Peterson SR, Kurimasa A, Oshimura M, Dynan WS, Bradbury EM, Chen DJ (1995) Loss of the catalytic subunit of the DNA-dependent protein kinase in DNA double-strand-break-repair mutant mammalian cells. Proc Natl Acad Sci USA 92:3171–3174

Rathmell WK, Chu G (1994a) A dna end-binding factor involved in double-strand break repair and v(d)j recombination. Mol Cell Biol 14:47414748

Rathmell WK, Chu G (1994b) Involvement of the Ku autoantigen in the cellular response to DNA double-strand breaks. Proc Natl Acad Sci USA 91(16):7623–7627

Roth DB, Menetski JP, Nakajima PB, Bosma MJ, Gellert M (1992) V(D)J recombination: broken DNA molecules with covalently sealed (hairpin) coding ends in scid mouse thymocytes. Cell 70:983–991

Schuler W, Weiler IJ, Schuler A, Phillips RA, Rosenberg N, Mak TW, Kearney JF, Perry RP, Bosma MJ (1986) Rearrangement of antigen receptor genes is defective in mice with severe combined immune deficiency. Cell 46:963–972

Schuler W, Ruetsch NR, Amsler M, Bosma MJ (1991) Coding joint formation of endogenous T cell receptor genes in lymphoid cells from scid mice: Unusual P-nucleotide additions in VJ-coding joints. Eur J Immunol 21(3):589–596

Sipley JD, Menninger JC, Hartley KO, Ward DC, Jackson SP, Anderson CW (1995) Gene for the catalytic subunit of the human DNA-activated protein kinase maps to the site of the XRCC7 gene on chromosome 8. Proc Natl Acad Sci USA (in press)

Smider V, Rathmell WK, Lieber MR, Chu G (1994) Restoration of X-ray resistance and V(D)J recombination in mutant-cells by Ku cDNA. Science 266(5183):288–291

Taccioli GE, Rathbun G, Oltz E, Stamato T, Jeggo PA, Alt FW (1993) Impairment of V(D)J recombination in double-strand break repair mutants. Science 260:207–210

Taccioli GE, Cheng H-L, Varghese AJ, Whitmore G, Alt FW (1994a) A DNA repair defect in Chinese hamster ovary cells affects V(D)J recombination similarly to the murine scid mutation. J Biol Chem 269:7439–7442

Taccioli TG, Gottlieb TM, Blunt T, Priestley A, Demengeot J, Mizuta R, Lehmann AR, Alt FW, Jackson SP, Jeggo PA (1994b) Ku80: product of the XRCC5 gene. Role in DNA repair and V(D)J recombination. Science 265:1442–1445

Thacker J, Wilkinson RE (1991) The genetic basis of resistance to ionising radiation damage in cultured mammalian cells. Mutat Res 254:135–142

Thompson LH, Jeggo PA (1995) Nomenclature of human genes involved in ionizing radiation sensitivity. Mutat Res:in press

van Gent DC, McBlane JF, Ramsden DA, Sadofsky MJ, Hesse JE, Gellert M (1995) Initiation of V(D)J recombination in a cell-free system. Cell 81:925–934

Weibezahn KF, Lohrer H, Herrlich P (1985) Double strand break repair and G2 block in Chinese hamster ovary cells and their radiosensitive mutants. Mutat Res 145:177–183

Whitmore GF, Varghese AJ, Gulyas S (1989) Cell cycle responses of two X-ray sensitive mutants defective in DNA repair. Int J Radiat Biol 56:657–665

Zhu C, Roth DB (1995) Characterization of coding ends in thymocutes of scid mice: implication for the mechanism of V(D)J recombination. Immunity 2:101–112

The DNA-Activated Protein Kinase – DNA-PK

Carl W. Anderson[1] and Timothy H. Carter[2]

1 Introduction:
The DNA-PK Lipid-like Kinase Superfamily

Recently, several mammalian and yeast genes have been identified that have carboxy-terminal domains (CTDs) related to the kinase domains of phosphatidylinositol kinases (PI3Ks) and appear to play roles in cell cycle control, DNA replication, recombination, and repair (Zakian 1995). The genes for this family of "lipid-like" kinases include *TOR1* (*DRR1*) and *TOR2* (*DRR2*) from *Saccharomyces cerevisiae* (Kunz et al. 1993; Helliwell et al. 1994), their mammalian homologue *FRAP* and *RAFT1* (also called *RAPT1*) (Brown et al. 1994; Chiu et al. 1994; Sabatini et al. 1994), *ATM*, the gene defective in patients with ataxia telangiectasia (Savitsky et al. 1995a,b), *MEI-41*, a *Drosophila* gene that is functionally similar to *ATM* (Hari et al. 1995), *MEC1* (*ESR1*) and its *S. pombe* counterpart *rad3* (Seaton et al. 1992; Kato and Ogawa 1994; Weinert et al. 1994), *TEL1*, a *S. cerevisiae* gene that controls telomere length (Greenwell et al. 1995; Morrow et al. 1995), and DNA-PK$_{cs}$ (*PRKDC*), the catalytic subunit of the DNA-activated protein kinase (Carter et al. 1990; Lees-Miller et al. 1990; Hartley et al. 1995; Poltoratsky et al. 1995). Tor1p, Tor2p, and FRAP control cell cycle progression and are targets of the immunosuppressive drug rapamycin. Mec1p/ESR1 is involved in the response to DNA damage and couples mitosis

[1]Biology Department, Brookhaven National Laboratory, Upton, NY 11973-5000, USA
[2]Department of Biological Sciences, St. John's University, New York, NY 11439, USA

to the completion of DNA synthesis. Cells with defects in the *ATM* gene are hyper-sensitive to ionizing radiation, and their ability to arrest cell cycle progression in response to DNA damage is defective or delayed. DNA-PK recently was found to be required for repairing DNA double-strand breaks and for the site-specific –V(D)J– recombination required to make functional immunoglobulin genes (BLUNT et al. 1995; KIRCHGESSNER et al. 1995; PETERSON et al. 1995b; ROTH et al. 1995). Together, these proteins form a distinct subgroup in the lipid kinase superfamily that can be distinguished from known PI3Ks (FRY 1994) by several properties (Fig. 1). Each encodes a protein that is substantially larger (>2500 amino acids) than the characterized lipid kinases; all are more closely related to each other in sequence and subdomain spacing than they are to the PI3Ks, and they share a segment of homology at their extreme carboxy-termini (Fig. 1, CR3) that is not present in PI3Ks (HARTLEY et al. 1995; POLTORATSKY et al. 1995). Thus, while these proteins may have evolved from lipid kinases, the properties of DNA-PK, the first of the family to be characterized biochemically, suggest that the physiological functions of these enzymes may

Lipid Kinase Superfamily Members

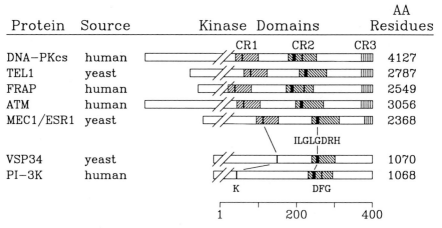

Fig. 1. Schematic diagram showing relationships between the kinase domains of PI kinase superfamily members. Polypeptides are represented by *open boxes*; the size of each is given *at the right*. Three regions of homology shared by members of the DNA-PK$_{cs}$ subgroup are indicated as *hatched segments*. Conserved region 1 (*CR1*) contains sequence corresponding to subdomain II of serine/threonine kinases including the conserved lysine (*vertical bar*) that interacts with the α and β phosphates of ATP (HANKS et al. 1988). Conserved region 2 contains sequence corresponding to S/T kinase subdomains VIb and VII; the highly conserved motif ILGLGDRH (*filled box*) probably corresponds to the HRDLKxxN motif of S/T kinases and contains the putative catalytic base (D) (HANKS and HUNTER 1995); only DNA-PK$_{cs}$, FRAP, and PI-3K have the conserved DFG motif (*vertical bar*) of S/T kinases. The conserved carboxy-terminal subdomain (CR3) of the DNA-PK$_{cs}$ subgroup is not present in PI kinases. Polypeptide sequences were obtained by translating the corresponding cDNA; protein sequences were partially aligned with the BLAST program (ALTSCHUL et al. 1990)

be to regulate cell cycle progression and DNA metabolism through phosphorylation of the concomitant protein substrates.

DNA-PK is a serine/threonine protein kinase that was shown to phosphorylate several DNA-binding proteins in vitro (for earlier reviews, see, e.g., CARTER and ANDERSON 1991; ANDERSON and LEES-MILLER 1992; ANDERSON 1993). DNA-PK is activated more than 700-fold by double-stranded DNA fragments and DNAs with single-to-double strand transitions, and binding to DNA is required to maintain the enzyme in an active conformation. These properties, together with recent genetic evidence, suggest that DNA-PK initiates signals in response to specific DNA structures (e.g., ends, nicks, gaps, hairpins, or bubbles) that accompany, control, or coordinate the steps required to join DNA fragments properly; it also may control the activities of enzymes involved in transcription, DNA replication, and other aspects of DNA repair. Here, we review the discovery and early history of DNA-PK and summarize its biochemical and physical properties.

2 Discovery of DNA-PK

DNA-PK was discovered as an activity in rabbit reticulocyte lysates and extracts of HeLa cells that promoted the transfer of ^{32}P-phosphate from γ-labeled ATP to endogenous protein substrates in response to exogenously added DNA fragments (WALKER et al. 1985; CARTER et al. 1988). Initially, it was uncertain that the activity was a protein kinase, because such labeling could have resulted from, for example, increased phosphate turnover due to activation of a phosphatase or an allosteric change in the substrate proteins upon binding DNA. However, characterization of one endogenous substrate, the heat shock protein hsp90 (LEES-MILLER and ANDERSON 1989a,b) and purification and characterization of the DNA-stimulated activity from cultured human HeLa cells clearly showed DNA-PK to be a previously undescribed serine/threonine protein kinase (CARTER et al. 1990; LEES-MILLER et al. 1990). DNA-PK is an abundant activity in cultured primate cells; therefore, one may wonder why it was not discovered earlier. One contributing factor is that protocols for preparing and purifying extracts often include steps for removing DNA; a second is that DNA primarily was thought of as a template or a substrate, and, at the time, little thought had been given to enzymes that could respond to DNA structures. Although 10 years have passed since the initial report of DNA-activated kinase activity, it still is one of the few known enzymes whose activity is regulated by DNA but does not appear to act directly on DNA.

Although DNA-PK is an abundant activity in cultured human cells derived from several tissues and in established cell lines from the Rhesus and African green monkeys, the activity is present at much lower (~50- to 100-fold) levels in rodent- and insect-derived cell lines (ANDERSON and LEES-MILLER 1992; FINNIE et al. 1995). A DNA-PK-like kinase activity also was observed in extracts prepared from the eggs and oocytes of clams (*Spisula*) and frogs (*Xenopus*) (WALKER et al. 1985; KLEINSCHMIDT and STEINBESSER 1991). Thus, DNA-PK or a DNA-PK-like protein kinase may be present in all metazoans. To our knowledge, a DNA-activated protein kinase has not been observed in extracts from single-cell organisms including the yeasts *S. cerevisiae* and *S. pombe*; however, the activity is difficult to detect in crude extracts unless it is very abundant. The recent discovery that DNA-PK belongs to a family of enzymes, several members of which (e.g., Tor1p, Tor2p, Mec1p, and Tel1p) are present in *S. cerevisiae*, suggests that a DNA-PK homologue also may be present in yeasts. However, whereas *S. cerevisiae* has a site-specific gene conversion mechanism that functions in mating-type switching, this mechanism differs from that of V(D)J recombination, and a DNA-activated kinase has not been shown to be involved either in it or in the pathway leading to the activation of mating type switching (HABER 1992; BARDWELL et al. 1994).

To date, DNA-PK has been purified and characterized only from human cells. In addition to HeLa cells, DNA-PK activity has been purified from human placenta (CHAN et al. 1995) and partially purified from Raji cells (IIJIMA et al. 1992; WATANABE et al. 1994b); the activity and polypeptide were detected in HL-60 cells (KONNO-SATO et al. 1993) and several other cultured human cell lines (ANDERSON and LEES-MILLER 1992). A nuclear casein kinase-like activity, termed NIII, was identified and partially purified from HeLa cells by FRIEDRICH and INGRAM (1989). NIII was chromatographically distinct from casein kinases I (CKI) and II (CKII), had a Stokes' radius consistent with an apparent molecular weight of 420kDa, used ATP in preference to GTP, and was inhibited at high ionic strengths. Although this enzyme was not shown to require DNA, based on these characteristics, it most probably is DNA-PK. To the extent examined, the properties of "DNA-PK" from each of these sources are indistinguishable.

3 Composition of Human DNA-PK

DNA-PK consists of a very large (~470 kDa) polypeptide, DNA-PK_{CS}, that contains the putative kinase catalytic domain (HARTLEY et al. 1995; POLTORATSKY et al. 1995) and a regulatory subunit(s) that targets DNA-PK_{CS} to DNA (Fig. 2). Although we did not demonstrate the presence of a regulatory/targeting subunit in our early studies, DVIR et al. (1992) and GOTTLIEB and JACKSON (1993) showed that the Ku autoantigen, a previously described DNA-binding protein (p70/p80), associates with DNA-PK_{CS}, targets it to DNA, and is required for the phosphorylation of several substrates (see below). Whether DNA-PK_{CS} associates with other targeting/regulatory subunits or can act as a protein kinase in their absence in vivo is unclear. Recent in vitro studies suggest that other DNA-binding proteins (e.g., RPA, GAL4, HSF1) may target DNA-PK_{CS} to DNA where it

The DNA-Activated Protein Kinase
DNA-PK

Signals   ## Targets

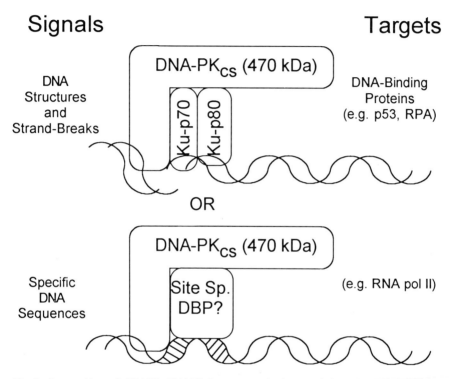

Fig. 2. Composition of DNA-PK. DNA-PK is composed of a catalytic polypeptide, DNA-PK$_{CS}$ (~470 kDa), and a DNA binding subunit, the Ku autoantigen, that targets DNA-PK$_{CS}$ to DNA structures. Ku is a heterodimer of ~70-kDa and ~80-kDa polypeptides. Other site specific DNA binding proteins (e.g., RPA, HSF1) also may target DNA-PK$_{CS}$ to DNA (see text)

becomes active (BRUSH et al. 1994; PETERSON et al. 1995b). The DNA-PK$_{CS}$ subunit also can be activated in vitro in the absence of known targeting subunits by certain oligonucleotide sequences (T.H. CARTER and N.P. MALIK, unpublished work). These results suggest that DNA-PK$_{CS}$ is activated through a direct interaction with DNA but that for most DNA structures, in the absence of a targeting subunit, this interaction is too weak to efficiently activate the kinase in vitro at the DNA concentrations normally used. The finding that rodent cells defective or deficient in either the Ku p80 subunit or DNA-PK$_{CS}$ have similar phenotypes (see below) suggests that Ku is a physiological targeting subunit for DNA-PK$_{CS}$, at least for DNA double-strand break repair and V(D)J recombination. Defects in either Ku-p80 or DNA-PK$_{CS}$ result in an increased sensitivity to ionizing radiation, defective repair of double-strand DNA breaks, and inability to recombine V(D)J substrates in the presence of exogenously supplied Rag1 and Rag2 (ROTH et al. 1995). Rag1 and Rag2 are required for V(D)J recombination but are not expressed in cell lines derived from fibroblasts and epithelial cells (see LEWIS 1994). In most biochemical studies of DNA-PK, Ku was present; thus, the results described below refer to the Ku:DNA-PK$_{CS}$ complex unless otherwise stated.

3.1 DNA-PK$_{CS}$

DNA-activated protein kinase activity and a very large, moderately abundant polypeptide from HeLa cell extracts were initially found to co-purify (CARTER et al. 1990; LEES-MILLER et al. 1990). Purification was monitored with four different protein substrates, hsp90, SV40 large T antigen (TAg), mouse p53, and beta-casein, and only one DNA-dependent kinase activity was observed in the extracts. The size of the large polypeptide was estimated from SDS-polyacrylamide gels to be 300–350 kDa; however, there is a notable lack of good molecular weight markers in this size range, and the size of the DNA-PK$_{CS}$ polypeptide based on the recently obtained sequence of its cDNA is close to 470 kDa (HARTLEY et al. 1995). The polypeptide has been given several names including p350, Prkdc, and DNA-PK$_{CS}$; however, we currently prefer DNA-PK$_{CS}$ to indicate that this subunit contains the putative catalytic site. The gene for the DNA-PK$_{CS}$ polypeptide was designated *PRKDC* (for protein kinase, DNA-activated, catalytic subunit) in accordance with the Genome Database nomenclature for other protein kinase genes.

Several early findings were consistent with assigning the catalytic site to the 470 kDa DNA-PK$_{CS}$ polypeptide (CARTER et al. 1990; LEES-MILLER et al. 1990; ANDERSON and LEES-MILLER 1992). DNA-PK$_{CS}$ binds ATP and can be labeled by the ATP analogues fluorosulfonylbenzoyladenosine (FSBA) and azido-ATP; FSBA inhibits DNA-PK kinase activity. Monoclonal antibodies specific for DNA-PK$_{CS}$ depleted DNA-dependent kinase activity from HeLa cell extracts, and one of these inhibited DNA-PK activity when added to partially purified enzyme. Alone, purified Ku antigen binds DNA but does not have homology with known protein kinases and does not exhibit protein kinase activity. Except for Ku, no other polypeptides have been obviously associated with DNA-PK$_{CS}$ during purification or are required for DNA-dependent protein kinase activity.

The ~13 500 nucleotide DNA-PK$_{CS}$ cDNA contains a single open reading frame over 12 300 nucleotides long that extends from the 5′ end to a TGA stop codon approximately 1600 nucleotides from its 3′ end (HARTLEY et al. 1995). Sixteen nucleotides before the 3′ polyA segment is an putative AAUAAA polyA addition signal. An in-frame ATG codon in a favorable Kozak consensus sequence is found near the 5′ end. The nascent DNA-PK$_{CS}$ polypeptide is predicted to consist of 4127 amino acids and to have a mass of 469 021 Da.

Although Northern blot analysis has revealed only a single mRNA species of ~12 kb, and fluorescent in situ hybridization (FISH) analysis identified only one chromosomal location for the DNA-PK$_{CS}$ gene, at present the possibility of alternate forms of the protein cannot be ruled out (see Note Added in Proof). Limited sequence analysis of gene fragments and comparisons with the cDNA suggests that the DNA-PK$_{CS}$ mRNA may be composed from nearly 100 exon sequences and that the size of the gene is approximately 180 kbp (SIPLEY et al. 1995). Several yeast artificial chromosomes (YAC) and cosmid clones that can restore DNA-PK activity to deficient cell lines have been identified (BLUNT et al. 1995; PETERSON et al. 1995a).

The DNA-PK$_{CS}$ sequence exhibits no significant homology with other proteins in Genbank (Release 89) except for homology between its carboxy-terminal lipid kinase domain and the carboxy-terminal kinase domains of other members of the lipid kinase superfamily. Thirty-four percent of the last 400 residues of DNA-PK$_{CS}$ and FRAP are identical while 47% are similar. The homologies with carboxy-terminal segments of TOR, ATM, MEC1, and TEL1 are similar; however, no obvious homology is found between segments of DNA-PK$_{CS}$ amino-terminal to the kinase domain and other members of this lipid kinase subgroup. The amino-terminal 3700 amino acid sequence of DNA-PK$_{CS}$ has few obvious sequence motifs. Interestingly, however, this segment does possess a leucine zipper element that may direct interactions with Ku or other proteins, and it contains numerous Ser-Gln motifs that represent potential DNA-PK autophosphorylation sites.

3.2 Ku Autoantigen

Ku is a heterodimer composed of two polypeptides, one of ~70 kDa and the other of ~80 kDa (p70/p80) (reviewed in REEVES 1992) that was originally identified as a human autoantigen associated with lupus and scleroderma overlap syndromes (MIMORI et al. 1981). It has been rediscovered many times and given several names (e.g., EBP-80, E$_1$BF, CHBF, CTCBF, NF IV, YPF1, TREF). Ku recognizes DNA ends and other DNA structures in the absence of DNA-PK$_{CS}$ (BLIER et al. 1993; FALZON et al. 1993). cDNAs for both human and mouse Ku polypeptides have been cloned and sequenced (CHAN et al. 1988; REEVES and STHOEGER 1989; YANEVA et al. 1989; MIMORI et al. 1990; FALZON and KUFF 1992; GRIFFITH et al. 1992), and the human genes recently were mapped to chromosomes 2 (p80) and 22 (p70) (CAI et al. 1994). Ku is moderately abundant in human cells where it is the major DNA-end binding activity, but like DNA-PK$_{CS}$, it appears to be much less abundant in rodent than in human cells (WANG et al. 1993). Ku may protect DNA ends from exonucleolytic degradation. Ku was reported to be associated with DNA replication complexes (CAO et al. 1994), and recombinant Ku was reported to have ATP-dependent helicase (human helicase II) and DNA-dependent ATPase activities (TUTEJA et al. 1994; VISHWANATHA et al. 1995). Ku or Ku-related proteins may act as transcription factors (MESSIER et al. 1993; HOFF et al. 1994; JACOBY and WENSINK 1994; ROBERTS et al. 1994), and the p80 subunit of Ku was reported to bind somatostatin and phosphatase 2A (LE ROMANCER et al. 1994). In vitro, Ku may translate along naked DNA molecules in an ATP-independent manner (DE VRIES et al. 1989); however, it is not known if it also can do so on chromatin or when it is associated with DNA-PK$_{CS}$. Ku proteins from nonmammalian species are less well characterized; however, Ku-related proteins have been described in *Drosophila* (JACOBY and WENSINK 1994) and in yeasts (FELDMANN and WINNACKER 1993).

Human HeLa cells contain about 400 000 molecules of Ku and perhaps ~50 000 to 100 000 molecules of DNA-PK$_{CS}$ (CARTER et al. 1990; ANDERSON and LEES-MILLER 1992). After exposure to γ radiation, the amount of the DNA-PK$_{CS}$ polypeptide and DNA-PK activity in extracts of HeLa cells did not change significantly within 24 h (T.H. CARTER and W. KAUFMANN, unpublished work; S.P. LEES-MILLER and C.W. ANDERSON, unpublished work). However, preliminary data from several laboratories suggest that exposing cells to DNA damaging agents may induce modifications to DNA-PK$_{CS}$ that alter its activity toward specific substrates. Exposure to elevated salt concentrations dissociates DNA-PK$_{CS}$ from Ku, and the rate of re-association is slow in the absence of DNA (SUWA et al. 1994), which suggests that active DNA-PK complexes may assemble at chromatin sites to which Ku has already bound. Both Ku polypeptides and DNA-PK$_{CS}$ have leucine zipper-like motifs that may function in assembling DNA-PK$_S$ and Ku or promoting interactions with other proteins.

4 Substrates and Substrate Specificity

In vitro DNA-PK phosphorylates a variety of nuclear DNA-binding proteins including the single-stranded DNA binding protein RPA, the tumor suppressor protein p53, several other transcription factors, the CTD of the large subunit of RNA polymerase II, and itself (Table 1). A few proteins that do not bind DNA, including hsp90, the microtubule-associated tau protein, casein, and phosvitin, also are phosphorylated by DNA-PK in vitro. In the few cases that have been studied, neither phosphorylation nor mutation of DNA-PK phosphorylation sites profoundly affected the activities of these substrates in vitro or in vivo. Thus, the biochemical consequences of phosphorylation by DNA-PK are largely unclear. Because many putative substrates are transcription factors and because RNA polymerase II itself is a substrate, it has been widely assumed that DNA-PK may regulate the expression of some RNA polymerase II transcripts. Recently, DNA-PK was shown to inhibit transcription from linear templates by RNA polymerase I in vitro (KUHN et al. 1995; LABHART 1995). DNA-PK did not phosphorylate pol I itself; rather, it appears to affect the activity of a transcription factor (SL1) required for pol I transcription. Phosphorylation by DNA-PK could block transcription of the ribosomal RNA precursors and other

Table 1. In vitro DNA-PK Substrates

Protein Substrate[a]	Reference
DNA-Binding Proteins	
DNA-PK$_{cs}$	CARTER et al. 1990; LEES-MILLER et al. 1990
Ku autoantigen, (p70 and p80)	LEES-MILLER et al. 1990
Replication Factor A (RPA)	ANDERSON 1993; BRUSH et al. 1994;
	PAN et al. 1994
[a] SV40 large Tumor Antigen (TAg)	CHEN et al. 1991
[a] Tumor suppressor protein p53	LEES-MILLER et al. 1990 1992;
	WANG and ECKHART 1992
[a] RNA Polymerase II CTD Domain	PETERSON et al. 1992; ANDERSON et al. 1995
[a] Serum Response factor (SRF)	LIU et al. 1993
[a] Transcription factor c-Jun	ANDERSON and LEES-MILLER 1992;
	ABATE et al. 1993; BANNISTER et al. 1993
[a] Transcription factor c-Fos	ANDERSON and LEES-MILLER 1992;
	ABATE et al. 1993
Transcription factor Oct1	ANDERSON and LEES-MILLER 1992;
	ANDERSON et al. 1995
Transcription factor Sp1	JACKSON et al. 1990
Transcription factor c-Myc	WATANABE et al. 1994b
Transcription factor CTF/NF-I	JACKSON et al. 1990
Transcription factor TFIID	ANDERSON and LEES-MILLER 1992
Chicken progesterone receptor	WEIGEL et al. 1992
Human estrogen receptor	ARNOLD et al. 1995
Topoisomerases I and II	ANDERSON and LEES-MILLER 1992
Xenopus Histone 2A.X	KLEINSCHMIDT and STEINBESSER 1991
Adenovirus 2 72-kDa DNA binding protein	ANDERSON and LEES-MILLER 1992
Bovine papillomavirus E2 protein	M. BOTCHAN, personal communication
Polyomavirus VP1	B. SCHAFFHAUSEN, personal communication
HMG1 and HMG2	WATANABE et al. 1994a
ERCC1	R. WOODS and C. W. ANDERSON, unpublished
E2F-1	E. HARLOW and C. W. ANDERSON, unpublished
Non-DNA-Binding Proteins	
[a] Heat shock protein 90 (hsp90)	WALKER et al. 1985;
	LEES-MILLER and ANDERSON 1989b
Microtubule-associated protein tau	WU et al. 1993
Casein	LEES-MILLER et al. 1990; CARTER et al. 1990
Phosvitin	LEES-MILLER et al. 1990; CARTER et al. 1990
Uncharacterized Substrates	
Unidentified HeLa 110-kDa polypeptide	LEES-MILLER et al. 1990
Unidentified HeLa 52-kDa polypeptide	LEES-MILLER et al. 1990
Unidentified HL-60 72 kDa polypeptide	KONNO-SATO et al. 1993

[a] Indicates substrates in which phosphorylation sites have been identified (see Table 2).

pol I transcripts in response to DNA damage or replication, but it remains to be determined if DNA-PK-mediated phosphorylation inhibits pol I transcription in vivo.

Another putative substrate that is receiving considerable attention is the single-stranded DNA binding protein RPA (also called HSSB). RPA is a heterotrimer with subunits of 70, 34, and 14 kDa. The 34 kDa subunit is phosphorylated at the G_1/S boundary and also becomes hyperphosphorylated after exposure to ionizing (Liu and Weaver 1993) and UV radiation (Carty et al. 1994). It was suggested that phosphorylation might act as a switch that regulates RPA's participation in chromosomal DNA replication versus repair processes. Although the 34 kDa subunit of RPA is phosphorylated by several kinases in vitro, DNA-PK was identified as the kinase responsible for hyperphosphorylation of RPA in cell extracts that support SV40 replication (Brush et al. 1994; Hendricksen et al., manuscript submitted). However, no difference was found in the ability of phosphorylated and unphosphorylated RPA to support SV40 replication or excision repair in vitro (Brush et al. 1994; Pan et al. 1995; Hendricksen et al., manuscript submitted). Although the function of RPA phosphorylation is unclear, there is some evidence that phosphorylation may modulate DNA replication, at least in vitro, by altering RPAs ability to interact with other replication proteins that also are phosphorylated by DNA-PK (Hendricksen et al., manuscript submitted). Boubnov and Weaver (1995) recently reported that RPA did not become hyperphosphorylated in murine *scid* cells exposed to ionizing radiation; this result strongly suggests that DNA-PK$_{cs}$ is responsible for the effect in vivo as well as in vitro.

The sites phosphorylated by DNA-PK in several substrates have been identified (Table 2); most are serines or threonines that are immediately followed in the amino acid sequence by glutamine (Anderson and Lees-Miller 1992; Lees-Miller et al. 1992; Anderson 1993). Hsp90 is a highly conserved heat-shock protein that is an abundant, predominantly cytoplasmic protein in unstressed cells. Two forms are present in most cells, alpha and beta, but only the alpha

Table 2. Protein phosphorylation sites recognized by human DNA-PK

Substrate protein[a]	DNA-PK site	Local amino acid sequence
Hsp90α (human)	Thr4	P E E **T Q** T Q D Q P M E E[13]
Hsp90α (human)	Thr6	P E E **T Q** T Q D Q P M E E E E[15]
SV40 Large Tumor Antigen	Ser120	E A T A D **S Q** H S T P P K K K[129]
SV40 Large Tumor Antigen	Ser665	E T G I D **S Q** S Q G S F Q A P[674]
SV40 Large Tumor Antigen	Ser667	G I D S Q **S Q** G S F Q A P Q S[676]
SV40 Large Tumor Antigen	Ser677	Q A P Q S **S Q** S V H D H N Q P[686]
c-Jun Transcription Factor (human)	Ser249	P I D M E **S Q** E R I K A E R K[258]
Serum Response Factor (human)	Ser435	V L N A F **S Q** A P S T M Q V S[444]
Serum Response Factor (human)	Ser446	M Q V S H **S Q** V Q E P G G V P[455]
p53 Tumor Suppressor (mouse)	Ser4	M E E **S Q** S D I S L E L P[13]
p53 Tumor Suppressor (mouse)	Ser15	L E L P L **S Q** E T F S G L W K[24]
p53 Tumor Suppressor (human)	Ser15	V E P P L **S Q** E T F S D L W K[24]
RNA Polymerase CTD Peptide	Ser7[b]	Y S P T S P **S** Y S P T S P S Y S

[a] Phosphorylation sites (boldface letters) were identified by: hsp90, Lees-Miller and Anderson (1989b); SV40 TAg, Chen et al. (1991); c-Jun, Bannister et al. (1993); serum response factor (SRF), Liu et al. (1993), p53 (Lees-Miller et al. (1992), and RNA polymerase CTD peptide (Anderson et al. 1995). Superscripted numbers at the right give the position of the associated amino acid in the nascent polypeptide.

[b] The seventh serine, as the repeat commonly is written, was phosphorylated in at least the first two repeats in the peptide.

form is phosphorylated by DNA-PK, which occurs at amino-terminal threonine residues that are not present in the beta form (LEES-MILLER and ANDERSON 1989b). However, in normal HeLa cells these sites do not appear to be phosphorylated at significant levels. SV40 T-antigen can be phosphorylated by DNA-PK on four serine residues, each of which is followed by glutamine (CHEN et al. 1991); at least two, Ser-120 and Ser-679, are phosphorylated in vivo. Mutating either of these sites affects T-antigen function, but the biochemical consequences of such mutations are not fully understood (MANFREDI and PRIVES 1994). Interestingly, T-antigen Ser-639, which is followed immediately by Gln-640, is phosphorylated in SV40-infected and transformed cells, but this site was not phosphorylated in vitro by DNA-PK (CHEN et al. 1991). Human serum response factor (SRF) is phosphorylated by DNA-PK at two serines (Ser-435 and Ser-446) in its transactivation domain, and peptide mapping indicated that both of these are phosphorylated in serum stimulated cells (LIU et al. 1993). Changing both serines to alanine in a GAL4-SRF fusion protein decreased the ability of this chimeric transcription factor to activate transcription of GAL4 reporter genes in transiently transfected cells. However, there is little evidence that DNA-PK plays a significant role in activating transcription through SRF in response to serum. c-Jun is phosphorylated by DNA-PK on Ser-249 (BANNISTER et al. 1993), but this site, while phosphorylated in vivo, also can be phosphorylated in vitro by casein kinase II. The human and murine p53 tumor suppressor proteins are phosphorylated by DNA-PK on serine 15 in the transactivation and mdm2-binding domain, and this residue also is phosphorylated in vivo (LEES-MILLER et al. 1992; WANG and ECKHART 1992; ULLRICH et al. 1993; MEEK 1994). Changing Ser-15 to alanine only marginally effected p53-mediated transactivation; however, this change slightly decreased the ability to arrest the growth of transiently transfected cells and significantly extended the half-life of the mutant protein (FISCELLA et al. 1993; MAYR et al. 1995). The site phosphorylated in a peptide containing four of the heptapeptide repeats from the CTD of the large subunit of RNA pol II recently was determined (ANDERSON et al. 1995). The CTD does not contain a serine-glutamine (-SQ-) nor a threonine-glutamine (-TQ-) pair, and, in this case, the phosphorylated serine is followed by tyrosine (Table 2).

DNA-PK substrate specificity also has been investigated using synthetic peptides (LEES-MILLER et al. 1992; ANDERSON et al. 1995). A 14-residue peptide containing two SQ sites from SV40 T-antigen was phosphorylated well by DNA-PK (CHEN et al. 1991) as were several peptides containing SQ sites from human p53 (LEES-MILLER et al. 1992). One of these, corresponding to sequence around serine 15 in conserved region 1 of the p53 transactivation domain, was modified systematically around the -SQ- site to determine the importance of sequence in substrate recognition. Substituting threonine for serine in the peptide EPPLSQEAFADLWKK yielded an equally good substrate; substituting glutamine or glutamic acid for the leucine before the phosphorylated serine reduced (improved) the apparent K_M about twofold; however, aspartic acid at this position did not do so. Interchanging the leucine and glutamine residues adjacent to the serine (to give -QSL-) had no appreciable effect on substrate activity; however, substituting glutamic acid, lysine, or tyrosine for glutamine dramatically decreased the rate of phosphorylation, and inverting the glutamine and the following glutamic acid (to give -LSEQ-) abolished substrate activity. Interestingly, tyrosine follows the serine in the CTD peptide that DNA-PK phosphorylates in vitro; thus, an adjacent glutamine is not an obligatory requirement for phosphorylation. Although the complete protein was not phosphorylated efficiently (IIJIMA et al. 1992), WATANABE et al. (1994b) recently reported that a peptide from the amino-terminal region of c-Myc, ELLPTPPLSPSRAFPE[69], was phosphorylated on serine and threonine, and a peptide from the carboxy-terminal region of the RB tumor suppressor protein, RPPTLSPIPHIPR[787], was phosphorylated on threonine by DNA-PK. They proposed that DNA-PK might function as a proline-directed protein kinase; however, the efficiency with which these peptides were phosphorylated was not determined. Many serine/threonine-containing peptides can be phosphorylated by DNA-PK if the peptide is provided at a sufficiently high concentration. The best peptide substrate for DNA-PK thus far described has the sequence PESQEAFADLWKK (ANDERSON et al. 1995).

In addition to the pol II CTD, several proteins are known that can be phosphorylated by DNA-PK at sites that do not involve the -SQ-/-TQ- motif (ANDERSON et al. 1995). Most have not been investigated, but several sites in recombinant rat c-*fos* were identified (ANDERSON and LEES-MILLER 1992). These findings suggest that sequence is not the sole determinant for substrate recognition by DNA-PK. A second element that clearly can be important in vitro is DNA binding. The rate of Sp1 phosphorylation was greatly stimulated by DNAs containing the Sp1 recognition site (GOTTLIEB and JACKSON 1993). This effect is not due to a change in Sp1 conformation since short GC box-containing oligonucleotides were not effective. Rather, an enhanced rate of phosphorylation resulted when DNA-PK and the substrate both bound to the same DNA molecule (ANDERSON and LEES-MILLER 1992; GOTTLIEB and JACKSON 1993). Thus, in some circumstances, DNA binding might be sufficient to drive phosphorylation at non-SQ/TQ sites (ANDERSON et al. 1995).

Although DNA-PK$_{cs}$ subfamily members are closely related to PI3Ks, it is not clear that they exhibit lipid kinase activity. Several lipid kinases were present in partially purified DNA-PK preparations; however, HARTLEY et al. (1995) did not find lipid kinase activity associated with highly purified DNA-PK$_{cs}$, either with or without DNA or Ku, using any of the phosphatidylinositol substrates presently available. Likewise, the TOR enzymes and FRAP have not been shown to possess lipid kinase activity (ZHENG et al. 1995). In contrast, POLTORATSKY et al. (1995) found that a PI4 kinase activity co-chromatographed with the peptide kinase activity of purified DNA-PK$_{cs}$. Further experiments will be required to resolve these differences and to determine if DNA-PK$_{cs}$ exhibits lipid kinase activity in vivo.

5 Activation by DNA

The purified Ku:DNA-PK$_{cs}$ complex is activated by any linear DNA fragment longer than about 12 bp (one turn of a helix), although slightly longer oligo-nucleotides (e.g., 18 bp or greater) activate more efficiently. Single-stranded deoxyhomopolymers competitively inhibit activation by dsDNA fragments; ssRNA, dsRNA, and RNA:DNA hybrids do not activate DNA-PK. Carter et al. (1990) showed that closed circular plasmids activated poorly compared with linear ones, and Anderson and Lees-Miller (1992) found that intact linear adeno-virus DNA, which has a peptide covalently attached at each 5′ termini, activated poorly compared to fragmented virus DNA. These findings suggested that activation might require binding to the ends of DNA fragments, a function that had been associated with the purified Ku protein (Mimori and Hardin 1986). However, more detailed studies showed that Ku could bind to nicked circular DNAs as well as closed circular DNAs containing single-to-double strand tran-sitions (e.g., closed hairpins, bubbles) (Blier et al. 1993; Falzon et al. 1993); subsequently, it was shown that these structures also activated DNA-PK (An-derson 1994; Morozov et al. 1994). These results suggest that in vivo DNA-PK might be activated by recombination intermediates, replication forks, and other structures as well as by nicks and DNA double-strand breaks. However, the extent of DNA-PK activation in vivo cannot be deduced from activity in cell extracts, and the fact that radioactive precursors produce DNA strand breaks complicates deductions based on analyses of putative substrates in cultured cells.

There have been persistent reports that Ku can bind DNA in a sequence-specific manner (e.g., Mitsis and Wensink 1989; Falzon and Kuff 1990; May et al. 1991; Okumura et al. 1994; Roberts et al. 1994; Genersch et al. 1995; Kim et al. 1995) and that DNA-PK can be activated by certain sequences better than by others (Carter et al. 1990; Lees-Miller et al. 1990; T.H. Carter, V.P. Poltoratsky and N.P. Malik, unpublished work). For DNA-PK activation, this is the case both for non-DNA binding substrates, such as casein and hsp90, and for at least some DNA-binding substrates including RPA, whose phos-phorylation is greatly stimulated by its own binding to DNA (M. Wold, unpub-lished) and the CTD of RNA polymerase II, which must be bound to the DNA template in order to be phosphorylated (Dvir et al. 1992). As noted above, sequence-specific activation of DNA-PK might be mediated by DNA-binding proteins other than Ku; for instance, HSF1, GAL4 (Peterson et al. 1995) and RPA (Brush et al. 1994). It remains to be seen if these factors, like Ku, serve as DNA-PK$_{cs}$ targeting subunits in vivo.

6 Regulation of DNA-PK Activity

The fact that DNA-PK is inactive unless physically associated with DNA means that its activity may be regulated in vivo not only by the availability and activity of regulatory/targeting subunits, but also by chromatin conformations that allow access to DNA, and by the presence of DNA structures to which its regulatory subunit(s) binds. Thus, while activation of DNA-PK due to the presence of appropriate DNA structures cannot readily be determined in vivo, by adding dsDNA fragments to crude or partially purified extracts, the amount of DNA-PK holoenzyme available for activation at the time of cell lysis can be assessed. It was found that when extracts were prepared from HeLa cells synchronized by exposure to hydroxyurea or nocodazole and then released, the amount of DNA-PK available for activation changed during the cell cycle, with the lowest during mitosis, followed by a tenfold increase during G_1 to a peak near the G_1/S border (SUN 1995). The increased activity could not be ascribed to changes in the amounts of the DNA-PK$_{cs}$ or Ku subunits and was, therefore, most likely due to post-translational modification of the enzyme. The amount of DNA-PK activity in extracts from metaphase cells was increased substantially by incubating the enzyme with alkaline phosphatase (SUN 1995). DNA-PK is inhibited by autophosphorylation (CARTER et al. 1990; LEES-MILLER et al. 1990); thus, the activity of DNA-PK, like the activities of several other cell kinases, may be modulated during the cell cycle by phosphorylation-dephosphorylation. Activation of DNA-PK at the G_1/S border would be consistent with a role in regulating DNA replication or repair through, for example, phosphorylation of RPA.

Recently, several laboratories have observed specific cleavage of DNA-PK$_{cs}$ in human cells treated to undergo apoptosis (CASCIOLA-ROSEN et al. 1995; Z.D. HAN, D. CHATTERJEE , T.H. CARTER, W. REEVES, J. WYCHE, and E.A. HENDRICKSON, manuscript submitted; Q. SONG, S.P. LEES-MILLER, S. KUMAR, N. ZHANG, D.W. CHAN, G.C.M. SMITH, S.P. JACKSON, E.S. ALNEMRI, G. LITMACK, and M.F. LAVIN, manuscript submitted; M. CARTY, personal communication). In HeLa cells exposed to UV-B, the majority of DNA-PK$_{cs}$ was cleaved to fragments of approximately 150 and 220 kDa, and the kinetics of cleavage were similar to that of poly(ADP-ribose) polymerase; neither subunit of Ku was cleaved (CASCIOLA-ROSEN et al. 1995). Cleavage was accompanied by a reduction in DNA-PK activity in cell extracts as measured by the phosphorylation of Sp1 (A. ROSEN, personal communication). DNA-PK$_{cs}$ was not cleaved in vitro by purified, recombinant human ICE, but incubation with apopain in the presence of double-stranded DNA, ATP, and Mg^{2+} resulted in cleavage between residues D^{2712} and N^{2713}. Omission of cofactors resulted in cleavage at additional sites generating smaller fragments. Cleavage of DNA-PK$_{cs}$ also has been observed in apoptotic human B lymphocytes after treatment with atoposide (Q. SONG et al., manuscript submitted), in Jurkat cells treated with anti-Fas antibody (K. MCCONNELL, personal communication), and in HL-60 cells treated with staurosporine (Z.D. HAN et al., manuscript submitted). A hallmark of apoptosis in most cells is extensive fragmentation of the genome (COLLINS 1991), and both poly(ADP-ribose) polymerase and DNA-PK are involved in the repair of DNA breaks. While it is uncertain whether the cleavage of these enzymes is an essential part of the apoptotic process, terminating processes that repair DNA breaks is consistent with this mechanism of programmed cell death.

7 Conclusions

The recent findings that certain ionizing radiation sensitive rodent cell lines defective for DNA double-strand break repair and V(D)J recombination also are deficient in, or express defective subunits of, DNA-PK strongly implicate this enzyme as an essential component in these processes. Cells in X-ray cross complementing group 5 (XRCC5) are deficient in DNA-PK activity and in the p80 subunit of Ku (GETTS and STAMATO 1994; SMIDER et al. 1994; TACCIOLI et al. 1994; FINNIE et al. 1995). Radiation resistance, DNA double-strand break repair capacity, V(D)J recombination ability, and DNA-PK activity all were restored by exogenously supplied DNA fragments or cDNAs that allow the Ku p80 subunit to be expressed (BOUBNOV et al. 1995; SMIDER et al. 1994; FINNIE et al. 1995). DNA-PK$_{cs}$ maps to the site of the XRCC7 gene (SIPLEY et al. 1995), and rodent cells from this complementation group, which includes the mouse *scid* (for severe combined immunodeficiency) and hamster V-3 cell lines (COLLINS 1993; THOMPSON and JEGGO 1995), are deficient in DNA-PK activity, expression of the DNA-PK$_{cs}$ polypeptide, DNA double-strand break repair, and in V(D)J recombination (BLUNT et al. 1995, KIRCHGESSNER et al. 1995; KOMATSU et al. 1995; PETERSON et al. 1995b), as is a recently described human tumor cell line, M059J (LEES-MILLER et al. 1995). DNA fragments that restored DNA-PK activity to these cell lines corrected the defects. Two additional rodent cell complementation groups also are defective for DNA double-strand break repair and V(D)J rejoining (COLLINS 1993); one of them may correspond to the p70 polypeptide of Ku (see THOMPSON and JEGGO 1995).

What role does DNA-PK play in DNA double-strand break repair and site-specific recombination? It seems likely that DNA-PK phosphorylates one or more of the components required for these processes, either to modulate a catalytic activity or to regulate assembly or disassembly of protein-DNA complexes (or both). We expect that at least one required component of each process will be a DNA-PK substrate. DNA-PK might also serve as a scaffold that holds individual protein components and the DNA ends in the proper spatial configuration for joining; providing such a scaffold might explain why the DNA-PK$_{cs}$ polypeptide is so large. That DNA-PK may function in regulating transcription, DNA replication, other DNA repair pathways and in cell cycle progression is strongly suggested by the multitude of putative DNA-PK substrates associated with these processes. The availability of cell lines defective in each of the DNA-PK components should facilitate the analysis of potential roles for DNA-PK in these processes.

Acknowledgments. Work performed in our laboratories is supported by the Office of Health and Environmental Research of the U. S. Department of Energy (C.W.A.) and by grants from the American Cancer Society and the Department of Health and Human Services (T.H.C). We gratefully acknowledge the contributions of many colleagues and collaborators.

Note Added in Proof. A comparison of the sequences of additional DNA-PK$_{cs}$ kinase domain cDNA clones with a sequence derived from HeLa cell cDNA clones (Genbank U34994) revealed a putative 93 bp exon that is present in the majority of DNA-PK$_{cs}$ mRNAs but absent in the former sequence (Poltoratsky et al. 1995; M.A. CONNELLY, H. ZHANG, J. KIELECZAWA, C.W. ANDERSON, Gene in press). This exon increases the predicted size of DNA-PK$_{cs}$ to 4127 amino acids (see Genbank).

References

Abate C, Baker SJ, Lees-Miller SP, Anderson CW, Marshak DR, Curran T (1993) Dimerization and DNA binding alter phosphorylation of Fos and Jun. Proc Natl Acad Sci USA 90: 6766–6770

Altschul SF, Gish W, Miller W, Myers EW, Lipman DJ (1990) Basic local alignment search tool. J Mol Biol 215:403–10

Anderson CW (1993) DNA damage and the DNA-activated protein kinase. Trends Biochem Sci 18:433–437

Anderson CW (1994) Protein kinases and the response to DNA damage. Semin Cell Biol 5:427–436

Anderson CW, Lees-Miller SP (1992) The human DNA-activated protein kinase, DNA-PK. Crit Rev Eukaryotic Gene Express 2: 283– 314

Anderson CW, Connelly MA, Lees-Miller SP, Lintott LG, Zhang H, Sipley JD, Sakaguchi K, Appella E (1995) Human DNA-activated protein kinase, DNA-PK: substrate specificity. In: Atassi MZ, Appella E (eds) Methods in protein structure analysis 1994. Plenum, New York pp 395–406

Arnold SF, Obourn JD, Yudt MR, Carter TH, Notides AC (1995) In vivo and in vitro phosphorylation of the human estrogen receptor. J Steroid Biochem Mol Biol 52:159–171

Bannister AJ, Gottlieb TM, Kouzarides T, Jackson SP (1993) c-Jun is phosphorylated by the DNA-dependent protein kinase in vitro definition of the minimal kinase recognition motif. Nucleic Acids Res 21:1289–1295

Bardwell L, Cook JG, Inouye CJ, Thorner J (1994) Signal propagation and regulation in the mating pheromone response pathway of the yeast Saccharomyces cerevisiae. Dev Biol 166:363–379

Blier PR, Griffith AJ, Craft J, Hardin JA (1993) Binding of Ku protein to DNA. Measurement of affinity for ends and demonstration of binding to nicks. J Biol Chem 268:7594–7601

Blunt T, Finnie NJ, Taccioli GE, Smith GCM, Demengeot J, Gottlieb TM, Mizuta R, Varghese AJ, Alt FW, Jeggo PA, Jackson SP (1995) Defective DNA-dependent protein kinase activity is linked to V(D)J recombination and DNA repair defects associated with the murine scid mutation. Cell 80:813–823

Boubnov NV, Weaver DT (1995) Scid cells are deficient in Ku and replication protein A phosphorylation by the DNA-dependent protein kinase. Mol Cell Biol 15:5700–5706

Boubnov NV, Hall KT, Wills Z, Lee SE, He DM, Benjamin DM, Pulaski CR, Band H, Reeves W, Hendrickson EA, Weaver DT (1995) Complementation of the ionizing radiation sensitivity, DNA end binding, and V(D)J recombination defects of double-strand break repair mutants by the p86 Ku autoantigen. Proc Natl Acad Sci USA 92:890–894

Brown EJ, Albers MW, Shin TB, Ichikawa K, Keith CT, Lane WS, Schreiber SL (1994) A mammalian protein targeted by G1-arresting rapamycin-receptor complex. Nature 369:756–758

Brush GS, Anderson CW, Kelly TJ (1994) The DNA-activated protein kinase is required for the phosphorylation of replication protein A during simian virus 40 DNA replication. Proc Natl Acad Sci USA 91:12520–12524

Cai Q-Q, Plet A, Imbert J, Lafage-Pochitaloff M, Cerdan C, Blanchard J-M (1994) Chromosomal location and expression of the genes coding for Ku p70 and p80 in human cell lines and normal tissues. Cytogenet Cell Genet 65:221–227

Cao QP, Pitt S, Leszyk J, Baril EF (1994) DNA-dependent ATPase from HeLa cells is related to human Ku autoantigen. Biochemistry 33:8548–8557

Carter TH, Anderson CW (1991) The DNA-activated protein kinase, DNA-PK. Prog Mol Subcell Biol 12:37–58

Carter TH, Kopman CR, James CBL (1988) DNA-stimulated protein phosphorylation in HeLa whole cell and nuclear extracts. Biochem Biophys Res Commun 157:535–540

Carter T, Vancurová I, Sun I, Lou W, DeLeon S (1990) A DNA- activated protein kinase from HeLa cell nuclei. Mol Cell Biol 10:6460–6471

Carty MP, Zernik–Kobak M, McGrath S, Dixon K (1994) UV light-induced DNA synthesis arrest in HeLa cells is associated with changes in phosphorylation of human single-stranded DNA-binding protein. EMBO J 13:2144–2123

Casciola-Rosen LA, Anhalt GJ, Rosen A (1995) DNA-dependent protein kinase is one of a subset of autoantigens specifically cleaved early during apoptosis. J Exp Med 182:1625–1634

Chan JYC, Lerman MI, Prabhakar BS, Isozaki O, Santisteban P, Notkins AL, Kohn LD (1988) Cloning and characterization of a cDNA that encodes a 70-kDa novel human thyroid autoantigen. J Biol Chem 264:3651–3654

Chan DW, Mody CH, Ting NSY, Lees-Miller SP (1995) Purification and characterization of the double-stranded DNA activated protein kinase, DNA-PK, from human placenta. Biochem Cell Biol 73(1112)67–73

Chen Y-R, Lees-Miller SP, Tegtmeyer P, Anderson CW (1991) The human DNA-activated protein kinase phosphorylates SV40 T-antigen at amino- and carboxy-terminal sites. J Virol 10:5131–5140

Chiu MI, Katz H, Berlin V (1994) RAPT1, a mammalian homolog of yeast Tor, interacts with FKBP12/repamycin complex. Proc Natl Acad Sci USA 91:12574–12578

Collins AR (1993) Mutant rodent cell lines sensitive to ultraviolet light, ionizing radiation and cross-linking agents: a comprehensive survey of genetic and biochemical characteristics. Mutat Res 293:99–118

Collins A (1991) Death by a thousand cuts. Curr Biol 1:140–142

de Vries E, van Driel W, Bergsma WG, Arnberg AC, van der Vliet PC (1989) HeLa nuclear protein recognizing DNA termini and translocating on DNA forming a regular DNA-multimeric protein complex. J Mol Biol 208:65–78

Dvir A, Peterson SR, Knuth MK, Lu H, Dynan WS (1992) Ku autoantigen is the regulatory component of a template-associated protein kinase that phosphorylates RNA polymerase II. Proc Natl Acad Sci USA 89:11920–11924

Falzon M, Kuff EL (1990) A variant binding sequence for transcription factor EBP-80 confers increased promoter activity on a retroviral long terminal repeat. J Biol Chem 265:13084–13090

Falzon M, Kuff EL (1992) The nucleotide sequence of a mouse cDNA encoding the 80 kDa subunit of the Ku (p70/p80) autoantigen. Nucleic Acids Res 20:3784

Falzon M, Fewell JW, Kuff EL (1993) EBP-80, a transcription factor closely resembling the human autoantigen Ku, recognizes single- to double-stranded transitions in DNA. J Biol Chem 268:10546–10552

Feldmann H, Winnacker EL (1993) A putative homologue of the human autoantigen Ku from Saccharomyces cerevisiae. J Biol Chem 268:12895–12900

Finnie NJ, Gottlieb TM, Blunt T, Jeggo PA, Jackson SP (1995) DNA- dependent protein kinase activity is absent in xrs-6 cells: implications for site-specific recombination and DNA double-strand break repair. Proc Natl Acad Sci USA 92:320–324

Fiscella M, Ullrich SJ, Zambrano N, Shields MT, Lin D, Lees-Miller SP, Anderson CW, Mercer WE, Appella E (1993) Mutation of the serine 15 phosphorylation site of human p53 reduces the ability of p53 to inhibit cell cycle progression. Oncogene 8:1519–1528

Friedrich TD, Ingram VM (1989) Identification of a novel casein kinase activity in HeLa cell nuclei. Biochem Biophys Acta 992:41–48

Fry MJ (1994) Structure and function of phosphoinositide 3-kinases. Biochem Biophys Acta 1226:237–268

Genersch E, Eckerskorn C, Lottspeich F, Herzog C, Kühn K, Pöschl E (1995) Purification of the sequence-specific transcription factor CTCBF, involved in the control of human collagen IV genes: subunits with homology to Ku antigen. EMBO J 14:791–800

Getts RC, Stamato TD (1994) Absence of a Ku-like DNA end binding activity in the xrs double-stranded DNA repair-deficient mutant. J Biol Chem 269:15981–15984

Peterson SR, Jesch SA, Chamberlin TN, Dvir A, Rabindran SK, Wu C, Dynan WS (1995a) Stimulation of the DNA-dependent protein kinase by RNA polymerase II transcriptional activator proteins. J Biol Chem 270:1449–1454

Peterson SR, Kurimasa A, Oshimura M, Dynan WS, Bradbury EM, Chen D (1995b) Loss of the catalytic subunit of the DNA-dependent protein kinase in DNA double-strand-break-repair mutant mammalian cells. Proc Natl Acad Sci USA 92:3171–3174

Poltoratsky VP, Lieber ML, Shi X, York JD, Carter TH (1995) Human DNA-PK is homologous to PI kinases. J Immunol 155:4529–4533

Reeves WH, Sthoeger ZM (1989) Molecular cloning of cDNA encoding the p70 (Ku) lupus auto-antigen. J Biol Chem 264:5047–5062

Reeves WH (1992) Antibodies to the p70/p80 (Ku) antigens in systemic lupus erythematosus. Rheumatic Dis Clinics of N America 18:391–414

Roberts MR, Young H, Fienberg A, Hunihan L, Ruddle FH (1994) A DNA-binding activity, TRAC, specific for the TRA element of the transferrin receptor gene copurifies with the Ku autoantigen. Proc Natl Acad Sci USA 91:6354–6358

Roth DB, Lindahl T, Gellert M (1995) How to make ends meet. Curr Biol 5:496–499

Sabatini DM, Erdjument-Bromage H, Lui M, Tempst P, Snyder SH (1994) RAFT1: A mammalian protein that binds to FKBP12 in a rapamycin-dependent fashion and is homologous to yeast TORs. Cell 78:35–43

Savitsky K, Bar-Shira A, Gilad S, Rotman G, Ziv Y, Vanagaite L, Tagle DA, Smith S, Uziel T, Sfez S, Ashkenazi M, Pecker I, Frydman M, Harnik R, Patanjali SR, Simmons A, Clines GA, Sartiel A, Gatti RA, Chessa L, Sanal O, Lavin MF, Jaspers NJ, Taylor AR, Arlett CF, Miki T, Weissman SM, Lovett M, Collins FS, Shiloh Y (1995a) A single ataxia-telangiectasia gene with a product similar to PI 3-kinase. Science 268:1749–1752

Savitsky K, Sfez S, Tagle D, Ziv Y, Sartiel A, Collins FS, Shiloh Y, Rotman G (1995b) The complete sequence of the coding region of the ATM gene reveals similarity to cell cycle regulators in different species. Hum Mol Genet 4:2025–2032

Seaton BL, Yucel J, Sunnerhagen P, Subramani S (1992) Isolation and characterization of the Schi-zosccharomyces pombe rad3 gene involved in the DNA damage and DNA synthesis checkpoints. Gene 119:83–89

Sipley JD, Menninger JC, Hartley KO, Ward DC, Jackson SP, Anderson CW (1995) Gene for the catalytic subunit of the human DNA-activated protein kinase maps to the site of the XRCC7 gene on chromosome 8. Proc Natl Acad Sci USA 92:7515–7519

Smider V, Rathmell WK, Lieber MR, Chu G (1994) Restoration of X-ray resistance and V(D)J rec-ombination in mutant cells by Ku cDNA. Science 266:288–291

Sun S-S (1995) Cell cycle regulation of the DNA-dependent protein kinase in HeLa cells. Doctoral Dissertation, St. John's University, Jamacia, NY

Suwa A, Hirakata M, Takeda Y, Jesch SA, Mimori T, Hardin JA (1994) DNA dependent protein kinase (Ku protein–p350 complex) assembles on double stranded DNA. Proc Natl Acad Sci USA 91:6904–6908

Taccioli GE, Gottlieb TM, Blunt T, Priestley A, Demengeot J, Mizuta R, Lehmann AR, Alt FW, Jackson SP, Jeggo PA (1994) Ku80: product of the XRCC5 gene and its role in DNA repair and V(D)J recombination. Science 265:1442–1445

Thompson LH, Jeggo PA (1995) Nomenclature of human genes involved in ionizing radiation sen-sitivity. Mutat Res 337:131–134

Tuteja N, Tuteja R, Ochem A, Taneja P, Huang NW, Simoncsits A, Susic S, Rahman K, Marusic L, Chen J, Zhang J, Wang S, Pongor S, Falaschi A (1994) Human DNA helicase II: a novel DNA unwinding enzyme identified as the Ku autoantigen. EMBO J. 13:4991–5001

Ullrich SJ, Sakaguchi K, Lees-Miller SP, Fiscella M, Mercer WE, Anderson CW, Appella E (1993) Phosphorylation at serine 15 and 392 in mutant p53s from human tumors is altered compared to wild-type p53. Proc Natl Acad Sci USA 90:5954–5958

Vishwanatha JK, Tauer TJ, Rhode SL III (1995) Characterization of the HeLa cell single-stranded DNA-dependent ATPase/DNA helicase II. Mol Cell Biochem 146:121–126

Walker, AI, Hunt T, Jackson RJ, Anderson CW (1985) Double- stranded DNA induces the phospho-rylation of several proteins including the 90 000 M_r heat-shock protein in animal cell extracts. EMBO J 4:139–145

Wang J, Chou C-H, Blankson J, Satoh M, Knuth MW, Eisenberg RA, Pisetsky DS, Reeves WH (1993) Murine monoclonal antibodies specific for conserved and non-conserved antigenic deter-minants of the human and murine Ku autoantigens. Mol Biol Rep 18:15–28

Wang Y, Eckhart W (1992) Phosphorylation sites in the amino- terminal region of mouse p53. Proc Natl Acad Sci USA 89: 4231– 4235

Watanabe F, Shirakawa H, Yoshida M, Tsukada K, Teraoka H (1994a) Stimulation of DNA-dependent protein kinase activity by high mobility group proteins 1 and 2. Biochem Biophys Res Comm 202: 736–742

Watanabe F, Teraoka H, Iijima S, Mimori T, Tsukada K (1994b) Molecular properties, substrate specificity and regulation of DNA-dependent kinase from Raji Burkett's lymphoma cells. Biochem Biophys Acta 1223:255–260

Weigel NL, Carter TH, Schrader WT, O'Malley BW (1992) Chicken progesterone receptor is phosphorylated by a DNA-dependent protein kinase during in vitro transcription assays. Mol Endocrinol 6:8–14

Weinert TA, Kiser GL, Hartwell LH (1994) Mitotic checkpoint genes in budding yeast and the dependence of mitosis on DNA replication and repair. Genes Dev 8:652–665

Wu JM, Chen Y, An S, Perruccio L, Abdel-Ghany M, Carter TH (1993) Phosphorylation of protein tau by double-stranded DNA-dependent protein kinase. Biochem Biophys Res Comm 193:13–18

Yaneva M, Wen J, Ayala A, Cook R (1989) cDNA-derived amino acid sequence of the 86-kDa subunit of the Ku antigen. J Biol Chem 264:13407–13411

Zakian VA (1995) ATM-related genes: what do they tell us about functions of the human gene? Cell 82:685–687

Zheng X-F, Fiorentino D, Chen J, Crabtree GR, Schreiber SL (1995) TOR kinase domains are required for two distinct functions, only one of which is inhibited by rapamycin. Cell 82:121–130

Role of the Ku Autoantigen
in V(D)J Recombination
and Double-Strand Break Repair

GILBERT CHU

1 Introduction

All cells have biochemical pathways for repairing DNA double-strand breaks induced by ionizing radiation (X-rays) and oxidative metabolism. Lymphoid cells also have a V(D)J recombination pathway for rearranging B cell immunoglobulin or T cell receptor genes (LEWIS 1994a). V(D)J recombination involves the cleavage of chromosomal DNA and subsequent resolution of the double-strand breaks. In the past few years, it has become clear that the two pathways share a number of common factors.

One of the factors involved in both V(D)J recombination and double-strand break repair (DSBR) was recently identified as the Ku autoantigen (GETTS and

Departments of Medicine and Biochemistry, Stanford University Medical Center, Stanford, CA 94305, USA

Stamato 1994; Rathmell and Chu 1994a,b; Smider et al. 1994; Taccioli et al. 1994). Ku was originally identified as the target antigen in a patient with the scleroderma-polymyositis overlap syndrome (Mimori et al. 1981). Because of its association with autoimmune disease, Ku has been extensively characterized over the last 15 years (Reeves 1992).

This paper will examine the biochemical role of the Ku autoantigen in V(D)J recombination and DSBR. To accomplish this, we will review what is known about the genetics of these pathways, the biochemistry of the Ku autoantigen, and the DNA intermediates in V(D)J recombination. We will consider a paradox raised by experiments studying the resolution of DNA hairpins in cells from the severe combined immunodeficient (scid) mouse. To resolve the paradox, we will propose a model for the role of Ku in V(D)J recombination and DSBR.

2 Genetics of V(D)J Recombination and DSBR

2.1 Recombination Signal Sequences and the RAG1 and RAG2 Genes

The germline loci of the immunoglobulin and T cell receptor genes contain multiple V, J, and in some cases D, coding elements. During the development of B and T cells, these elements undergo V(D)J recombination, which involves the rearrangement of V, D, and J elements to form functional genes (Lewis 1994a). The multiplicity of elements generates diversity in the expressed proteins. Rearrangements are targeted to recombination signal sequences (RSS), which consist of conserved heptamer and nonamer sequences separated by a spacer of 12 or 23 base pairs. Coding elements are fused at coding joints to form rearranged genes capable of encoding functional proteins. The coding joints are formed with deletions or insertions that generate additional diversity. Signal elements are fused at signal joints, which are formed precisely with exact conservation of the RSS.

V(D)J recombination is initiated by the combined action of the recombination activating genes, RAG1 and RAG2, which are expressed specifically in lymphoid cells. Cotransfection of RAG1 and RAG2 confers V(D)J recombination activity to nonlymphoid cells (Oettinger et al. 1990; Schatz et al. 1989). Thus, once V(D)J recombination is initiated by RAG1 and RAG2, general factors present in all cells will complete the recombination reaction. These general factors also have a second function in DSBR, as discussed below.

2.2 The Scid Mouse Is Defective in Both V(D)J Recombination and DSBR

The scid mouse lacks mature T and B cells and is highly susceptible to the development of T cell lymphomas (BOSMA and CARROLL 1991). The failure to develop a competent immune system was explained by the discovery that scid cells cannot form coding joints during V(D)J recombination (LIEBER et al. 1988). By contrast, scid cells can form signal joints at a normal rate, although only 50% of the joints are precise.

In addition to a lymphoid-specific defect, scid cells were found to have a general defect in DSBR affecting all tissues and causing hypersensitivity to ionizing radiation due to a deficiency in repairing DNA double-strand breaks (BIEDERMANN et al. 1991; FULOP and PHILLIPS 1990; HENDRICKSON et al. 1991). This was the first evidence that V(D)J recombination and DSBR utilize common factors present in all tissues.

2.3 Multiple Cell Lines Defective for Both V(D)J Recombination and DSBR

Ionizing radiation produces several different DNA lesions, including base damage, single-strand breaks, and double-strand breaks. To search for the genetic basis of resistance to ionizing radiation, a number of easily cultured hamster cell lines have been developed by mutagenesis and screening for hypersensitivity to ionizing radiation (JEGGO 1990). Cell fusion experiments show that these cell lines fall into at least ten genetic complementation groups. The corresponding genes are designated XRCC1, XRCC2, etc., for X-ray cross-complementing, since early efforts were aimed at cloning the genes by cross-complementation of the X-ray-sensitive hamster cells with human DNA.

The discovery that scid cells are sensitive to X-rays raised the possibility that other X-ray sensitive cells might also be defective in V(D)J recombination. Therefore, cell lines were screened by cotransfection of RAG1, RAG2, and an extrachromosomal V(D)J recombination substrate. Cell lines from each of three complementation groups with defects in DSBR also proved to be defective in V(D)J recombination (LEE et al. 1995; PERGOLA et al. 1993; TACCIOLI et al. 1993). Complementation group 7 (which includes scid) was defective for coding joint but not signal joint formation. Complementation groups 4 and 5 were defective for both coding and signal joint formation. The residual recombination events recovered from mutant cells were characterized by abnormally large nucleotide deletions in the coding joints in group 7 or both coding and signal joints in groups 4, and 5. Thus, at least three gene products are involved in a pathway common to both V(D)J recombination and DSBR. The genetics of V(D)J recombination are summarized in Fig. 1.

A possible explanation for the large deletions during the joining reaction was that the cells might be defective in a protein that bound and protected

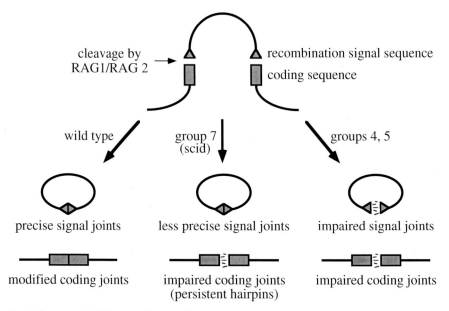

Fig. 1. Genetics of V(D)J recombination. The RAG1 and RAG2 genes induce cleavage adjacent to RSS. In wild-type cells, recombination then produces precise signal joints and modified coding joints. In cells from X-ray sensitivity complementation group 7, which corresponds to the scid defect, coding joints are severely impaired while signal joints are only mildly impaired. The impairment in coding joints is accompanied by the persistence of hairpin ends. In cells from groups 4, and 5, both coding and signal joints are severely impaired

the DNA ends from nuclease degradation. This possibility was confirmed by discovery of the normal role of the Ku autoantigen in cells.

3 The Ku Autoantigen

Ku is the target antigen in patients with several autoimmune diseases, including scleroderma-polymyositis overlap syndrome, sytemic lupus erythematosus, Grave's disease, and Sjögren's syndrome (REEVES 1992). Ku is a highly stable heterodimer of 70 kDa and 86 kDa polypeptides (Ku70 and Ku80) (MIMORI et al. 1986). The cDNAs for both subunits have been cloned (MIMORI et al. 1990; REEVES and STHOEGER 1989; YANEVA et al. 1989). Ku is localized to the nucleus and moderately abundant, with about 200 000 to 400 000 molecules present in each cell.

Ku has an interesting specificity for its DNA substrates. It does not bind to single-stranded DNA ends, but binds tightly to double-stranded ends, having equal affinity for 5' overhanging, 3' overhanging, and blunt ends (MIMORI and HARDIN 1986). Ku also binds to DNA nicks (BLIER et al. 1992) and to DNA ending in stem loop structures (FALZON et al. 1993). This spectrum of DNA binding

DNA substrates   Ku binding

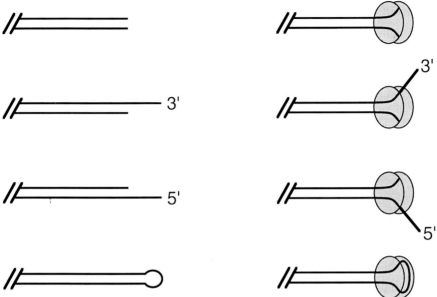

Fig. 2. Model for Ku DEB activity. Ku binds to a number of DNA substrates, including DNA with blunt ends, 5' or 3' overhanging ends, or stem loop structures. Ku may recognize all of these structures by binding to the forked DNA structure formed at the transition of double-stranded DNA to two single strands

activity can be explained by a model in which Ku recognizes transitions between double-stranded DNA and two single strands, as depicted in Fig. 2. Once Ku binds to DNA ends, it is capable of translocating along the DNA, so that three or more molecules of Ku can become bound to a single DNA fragment (PAILLARD and STRAUSS 1991).

Ku is conserved both functionally and structurally across a broad range of eukaryotes. The fruit fly *Drosophila melanogaster* (BEALL et al. 1994; JACOBY and WENSINK 1994) and yeast *Saccharomyces cerevisiae* (FELDMANN and WINNACKER 1993) contain DNA end-binding (DEB) activities arising from heterodimeric proteins of similar molecular weights as mammalian Ku. In both cases, the gene for the smaller (70 kDa) subunit was cloned and found to be homologous to mammalian Ku70. The genome project for the nematode *Caenorhabditis elegans* has revealed an open reading frame encoding a polypeptide homologous to mammalian Ku80.

Ku is the regulatory subunit of an unusual enzyme, DNA-dependent protein kinase (DNA-PK), which also includes an enormous 460 kDa catalytic subunit (DNA-PK$_{cs}$). DNA-PK is normally inactive in free solution, but when it assembles on a DNA substrate recognized by Ku, DNA-PK is activated for its kinase activity (GOTTLIEB and JACKSON 1993). Although the in vivo substrates of DNA-PK remain to be defined, potential substrates have been identified in vitro, including Ku70 and Ku80, tumor suppressor protein p53, the C-terminal domain of RNA polymerase II, replication protein A (RPA), topoisomerases I and II, serum response factor (SRF), SV40 T antigen, *Xenopus* histone protein 2A.X, and the transcription factors c-*myc*, c-*fos*, c-*jun*, Sp1, TFIID and Oct-1 (ANDERSON 1993). DNA-PK will phosphorylate Sp1 when it is bound to the same DNA molecule, even when separated by several kilobases, but will not act on Sp1 bound to a different DNA molecule (GOTTLIEB and JACKSON 1993).

Ku undergoes post-translational modification. When Ku is autophosphorylated by DNA-PK, it acquires an ATPase activity (CAO et al. 1994), making Ku an ATP dependent helicase (TUTEJA et al. 1994). When Ku binds to one of its substrates, it recruits DNA-PK$_{CS}$ and then is autophosphorylated to become an active helicase. Ku may also be processed by proteolytic cleavage of the Ku80 subunit (PAILLARD and STRAUSS 1993).

4 Identification of Genes Involved in both V(D)J Recombination and DSBR

4.1 DEB Activity in Mutant and Wild-Type Cells

DEB activity can be detected in cells by an electrophoretic mobility shift assay (EMSA) (RATHMELL and CHU 1994a). Extracts of the cells are incubated with a labeled linear fragment of DNA in the presence of unlabeled circular plasmid DNA to mask the effect of nonspecific DNA binding proteins. The reaction mixture is resolved by nondenaturing polyacrylamide gel electrophoresis, and DEB activity is detected as an upward shift in electrophoretic mobility of the DNA probe.

DEB activity was found to be present in cells from yeast and from a number of tissues from humans and rodents (Table 1) (RATHMELL and CHU 1994a). It was expressed normally in cell lines from groups 1, 4, and 7 (including scid cells), in an ataxia telangiectasia cell line, two bleomycin-sensitive cell lines, and a mutant X-ray sensitive lymphoblast cell line. DEB activity was notably absent in three cell lines from X-ray complementation group 5, XR-V15B, XR-V9B, and xrs5. Azacytidine will induce xrs5 cells to revert to X-ray resistance at a frequency of about 1%. Significantly, 23 independent clones selected for X-ray resistance all showed restored DEB activity, further supporting a role for DEB activity in X-ray resistance.

Table 1. DEB activity in wild-type and mutant cell lines

Cell type	Cell lines	Resistance to IR	Complement group	DEB activity
Yeast				
S. cerevisiae	hdf1/pHDF1	+		+
S. cerevisiae	hdf1	+		−
Human				
Fibroblast (wild-type)	IMR-90	+		+
Ataxia telangiectasia fibroblast	AT5BI	−		+
Mouse				
Skin fibroblast (wild-type)	C.B-17	+		+
Skin fibroblast	scid	−	7	+
Chinese hamster				
Lung fibroblast (wild-type)	V79, V79B	+		+
Lung fibroblast	XR-V15B, XR-V9B	−	5	−
Ovary (wild-type)	AA8	+		+
Ovary	EM9	−	1	+
Ovary	XR-1	−	4	+
Ovary	V3	−	7	+
Ovary	xrs5, xrs6, sxi2, sxi3	−	5	−
Ovary	xrs5(rev)[a]	+		+

DEB activity was measured in yeast, human, and rodent cells by the gel mobility shift assay. DEB activity is specifically absent in complementation group 5. DEB activity is present in several other sensitive cell lines, including ataxia telangiectasia cells and cells from complementation groups 1, 4, and 7 (including scid). The table compiles data from (BOUBNOV et al. 1995; LEE et al. 1995) for sxi cells, from (GETTS and STAMATO 1994) for xrs6 cells, and from (RATHMELL and CHU 1994a) for the remaining cell lines. The HDF1 gene encodes a 70 kDa subunit of a DEB activity in yeast that is homologous to mammalian Ku70 (FELDMANN and WINNACKER 1993). The deletion mutant hdf1 is resistant to ionizing radiation (S. CHENG and G. CHU, unpublished results), suggesting that the dominant mechanisms for X-ray resistance differ in yeast and mammals.
[a] xrs5(rev) represents pooled xrs5 cells selected for X-ray resistance after azacytidine treatment.

4.2 DEB Activity Is Due to Ku Autoantigen

DEB activity and Ku share many similarities, including nuclear localization, abundance in the cell, and magnesium-dependent DNA binding (RATHMELL and CHU 1994b). Most striking is the absolute concordance of DNA substrates. DEB activity and Ku bind to M13 circular virion DNA (presumably via binding to hairpin loops in M13), double-stranded DNA ends with 5′, 3′, or blunt ends, but not to the single-stranded DNA ends in poly(dA) and poly(dT). Three or more Ku molecules can load and bind to a single DNA fragment, producing a ladder of mobility shifts similar to that seen with DEB activity.

DEB activity and Ku are antigenically similar (GETTS and STAMATO 1994; RATH-MELL and CHU 1994b). When added to binding reactions with hamster and human extracts, Ku antisera from two different autoimmune patients produce a supershift in mobility of the DEB protein–DNA complex. When used to probe immunoblots, these antisera detect a 70 kDa polypeptide in hamster extracts that cofractionates with DEB activity on heparin agarose. The 70 kDa polypeptide was absent or only barely detectable in three hamster cell lines from group 5, but present in group 4, group 7, and wild-type cell lines (RATHMELL and CHU 1994b). Thus Ku70 polypeptide is deficient in group 5 cells.

Despite this observation, Ku70 does not define the genetic defect in group 5. Chromosome mapping studies assigned the Ku70 gene to human chromosome 22q13 and the Ku80 gene to chromosome 2q33-35 (CAI et al. 1994; BOUBNOV et al. 1995). X-ray resistance in group 5 cells was partially restored by human chromosome region 2q35 (CHEN et al. 1994; HAFEZPARAST et al. 1993). This raised the possibility that the primary defect in group 5 resides in the gene for Ku80.

Direct evidence that the XRCC5 gene was identical to the Ku80 gene was provided by DNA transfection experiments. Thus, transfection of group 5 cells with human Ku80 but not Ku70 led to partial restoration of DEB activity, X-ray resistance, and V(D)J recombination (SMIDER et al. 1994; TACCIOLI et al. 1994). Full restoration was obtained by the isolation and transfection of the hamster Ku80 cDNA. In addition, Ku70 protein levels were restored, suggesting that the Ku70 polypeptide is stabilized by the presence of normal Ku80 polypeptide (ERRAMI et al. 1995). Finally, Ku80 cDNA from XR-V15B and XR-V9B cells contained in-frame deletions of 46 and 84 amino acids, proving that mutations in the Ku80 gene are responsible for the phenotype of group 5 cells (ERRAMI et al. 1995).

The finding that the Ku80 gene is identical to the XRCC5 gene immediately raised the possibility that DNA-PK$_{cs}$ might be encoded by another XRCC gene involved in DSBR. Yeast artificial chromosomes carrying the DNA-PK$_{cs}$ gene will rescue group 7 cells, making DNA-PK$_{cs}$ a candidate for XRCC7 (BLUNT et al. 1995; KIRSCHGESSNER et al. 1995). Recently, a novel cDNA has been isolated as a candidate for XRCC4 by complementation of XR-1 cells (see LI et al. 1995; Z. LI and F. ALT, this volume). No Ku70 mutant has yet been confirmed.

These experiments demonstrate that V(D)J recombination and DSBR include an overlapping pathway. Furthermore, at least four different genes act in the overlapping pathway (Table 2). Together with recent progress in identifying the enzymatic activities of RAG1 and RAG2, it is now possible to construct a model for the biochemistry of V(D)J recombination. Before introducing the model, it will be necessary to examine the DNA intermediates in V(D)J recombination.

Table 2. Genes involved in V(D)J recombination

Gene	Mutant cell lines	IR	V(D)J activity	
			Coding joints	Signal joints
Specific for V(D)J				
RAG1	KO mouse	R	No cleavage	No cleavage
RAG2	KO mouse	R	No cleavage	No cleavage
TdT	KO mouse	R	No N-addition	Normal
General for DSBR				
XR-1 (XRCC4)	XR-1	S	Deficient	Deficient
Ku80 (XRCC5)	xrs5, xrs6, sxi2, sxi3, XR-V15B, XR-V9B	S	Deficient	Deficient
Ku70 (XRCC6)	sxi1 ?	S ?	Deficient ?	Deficient ?
DNA-PK$_{cs}$ (XRCC7)	scid mouse, V-3	S	Deficient	Normal rate, partial fidelity

The genes involved in V(D)J recombination fall into two classes: those specific for V(D)J recombination and those general for DSBR. Some of the mutant cell lines were generated from mice with the scid mutation or with targeted knockout (KO) of the RAG1, RAG2, and TdT genes. The remaining cell lines were generated from mutagenesis of Chinese hamster cells. Cell lines mutant for genes specific for V(D)J recombination are resistant (R) to ionizing radiation (IR), whereas the cell lines mutant for genes general for DSBR are sensitive to ionizing radiation and fall into X-ray complementation groups 4, 5, or 7. Group 6 has been reserved for Ku70 (XRCC6), but no cell line has been found yet for this group.

5 DNA Intermediates in V(D)J Recombination

5.1 Formation of Broken DNA Ends

The association between DSBR and V(D)J recombination suggests that V(D)J joining involves DNA intermediates in which both strands have been broken. Such broken molecules have been observed directly. Thymocytes from newborn mice actively rearrange the T cell receptor locus and contain broken DNA with blunt signal ends (ROTH et al. 1992). On the other hand, coding ends are not detectable, even though a primary double-strand break should liberate one coding end for each signal end.

In fact, coding and signal joints are formed differently. Signal joints are formed without the addition or loss of nucleotides. Coding joints often contain either deletions or short insertions of extra nucleotides not present in the germline DNA. In some cases, the insertions have short palindromic sequences derived from one of the coding ends (LAFAILLE et al. 1989; McCORMACK et al. 1989). These P (palindromic) insertions are potentially explained by a model in which coding ends are created by the formation of a hairpin intermediate (LIEBER 1991). If the hairpin is nicked at a position away from the tip, a palindromic sequence appears in the completed coding joint. The first direct evidence for hairpin ends was found in scid thymocytes (ROTH et al. 1992). Hairpin ends were not found in wild-type thymocytes, suggesting that coding joints are formed much more rapidly than signal joints. Thus, the scid defect appears to disrupt hairpin processing, allowing hairpin ends to accumulate to detectable levels in scid but not wild-type thymocytes.

Hairpin ends are created as part of the V(D)J cleavage reaction (see Fig. 3). In an experimental tour de force, purified recombinant RAG1 and RAG2 proteins were recently shown to directly catalyze a double-strand break in DNA molecules containing a recombination signal sequence (van Gent et al. 1995). The reaction is absolutely dependent on the RSS and both RAG1 and RAG2. Cleavage occurs in a two-step reaction in which RAG1 and RAG2 first nick the DNA 5′ to the RSS; after a time delay, RAG1 and RAG2 catalyze a nucleophilic attack by the 3′ OH of the opposite strand (McBlane et al. 1995). This second step generates a blunt signal end and a hairpin coding end.

The coding ends are subjected to further modification by nucleases (N deletion) or by the nontemplated addition of nucleotides (N insertion). Most N addition occurs by 3′ addition catalyzed by terminal deoxynucleotidyl transferase (TdT), which is expressed only in lymphoid cells. A low rate of N insertion occurs in all cells, perhaps by the capture of free nucleotides or oligonucleotides (Roth et al. 1989).

5.2 Joining Reactions

The rejoining reaction in V(D)J recombination does not require extensive homology in the recombining DNA. V(D)J recombination will still occur in extrachromosomal substrates with homopolymeric coding sequences that do not permit homologous pairing (Boubnov et al. 1993).

On the other hand, V(D)J recombination preferentially utilizes short stretches of homology if they are present. When nonlymphoid cells are cotransfected with RAG1, RAG2, and an extrachromosomal substrate, the majority of coding joints are formed at positions containing short homologies of 1–5 bp. In extrachromosomal substrates constructed with 4 bp of homology at the two coding ends, there was a strong bias towards the coding joint formed by homology alignment of the 4 bp repeats (Gerstein and Lieber 1993).

In lymphoid cells, homology pairing may be obscured by TdT activity, since N insertion can include homologous nucleotides that would disappear once the joint is made. In lymphoid cells from TdT knockout mice, N insertion is virtually eliminated (Gilfillan et al. 1993; Komori et al. 1993). In the absence of TdT, homology alignment occurs in 75% of the coding joints. Thus, the use of homology is preferred, but not essential for the formation of coding joints. Homology affects the distribution of coding joints but not the overall efficiency of the reaction.

General DSBR has been studied by transfecting linearized SV40 DNA into nonlymphoid cells (Roth et al. 1985; Roth and Wilson 1986). These plasmids did not contain RSS and were linearized by cutting the SV40 genome in a

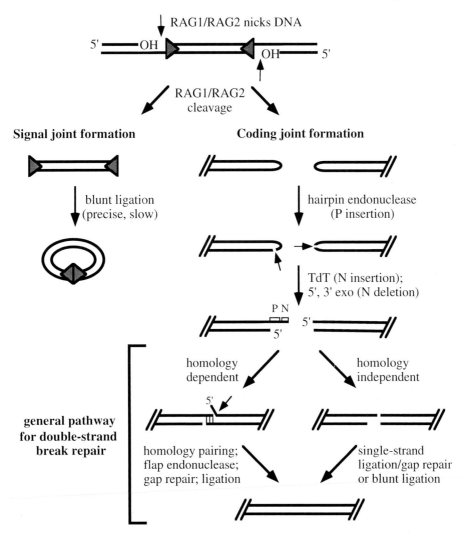

Fig. 3. DNA intermediates during V(D)J recombination. RAG1 and RAG2 recognize RSS (*shaded triangles*) and nick the adjacent DNA, leaving a 3' hydroxyl group. RAG1 and RAG2 then catalyze a nucleophilic attack by the 3' OH of the opposite strand to complete the cleavage reaction, leaving a blunt signal end and a hairpin coding end. A hairpin endonuclease then opens the hairpin symmetrically to leave a blunt end, or asymmetrically to leave an overhanging end that potentially leads to a palindromic sequence at the coding joint (P insertion). Nucleotides may be added (N insertion) by terminal deoxynucleotidyl transferase (TdT) or deleted (N deletion) by the activity of exonucleases. The ends are joined by homology-dependent or homology-independent mechanisms. Homology-dependent joining may involve pairing in regions of microhomology, removal of unpaired end fragments by a flap endonuclease, repair of gaps by DNA polymerase, and ligation of the nicks. Homology-independent joining involves either single-strand ligation and gap repair or blunt ligation

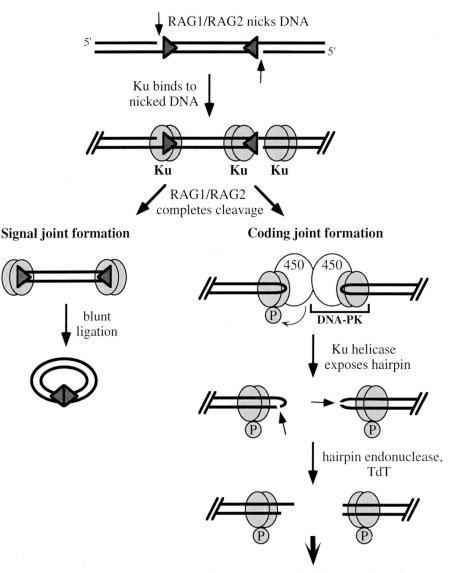

Fig. 4. Model for Ku in V(D)J recombination. After RAG1 and RAG2 nick the DNA at the RSS (*shaded triangles*), Ku (*shaded dimer*) binds to the nicked DNA. Therefore, when cleavage is completed, Ku is immediately loaded onto the newly created signal and coding ends. The signal joint is dependent on Ku but not DNA-PK$_{cs}$, the 460 kDa catalytic subunit of DNA-PK. The coding joint is formed by processing of the hairpin ends. DNA-PK is assembled by recruitment of DNA-PK$_{cs}$ to the hairpin end. DNA-PK$_{cs}$ has an alignment function to bring the DNA ends in proximity to each other. DNA-PK autophosphorylates Ku (*encircled Ps*), conferring ATP dependent helicase activity to Ku. The helicase activity unwinds the hairpin, allowing hairpin endonuclease to nick the hairpin. Nucleotides may be added by TdT or deleted by exonucleases. The free ends are then joined by the general pathway for DSBR

nonessential intron, leaving mismatched ends. As for the case of V(D)J recombination, joining occurred by both homology-dependent and homology-independent mechanisms. Homology-independent joining appears to utilize single-strand ligation preferentially over blunt ligation (ROTH and WILSON 1986). This general pathway for DSBR is shown at the bottom of Fig. 3.

In summary, the joining reactions in V(D)J recombination and general DSBR share a common pathway. Both reactions are mediated by the same proteins, namely, Ku, DNA-PK$_{CS}$, and the XR-1 gene product. Both reactions involve similar DNA intermediates with joining reactions both dependent and independent of microhomology. Pairing by microhomology suggests a mechanism for unwinding the DNA ends, perhaps catalyzed by Ku helicase activity.

6 A Paradox of Hairpin Processing in Scid Cells

The persistence of hairpin ends in scid thymocytes suggests that the scid mutation leads to defective processing of hairpin intermediates during V(D)J recombination. However, when cells were transfected with linearized plasmid DNA carrying hairpin ends, the rejoining reaction occurred with the same efficiency in scid and wild-type pre-B cells (LEWIS 1994b). The joint sequences showed patterns of deletion, P insertion, N insertion, and microhomology usage that were similar in the scid and wild-type cells and similar to the coding joints generated by bona fide V(D)J recombination. Thus we are faced with an apparent experimental paradox: scid cells cannot join hairpin ends created during V(D)J recombination, but can process and join hairpins introduced into the cells by transfection.

7 Solution to the Scid Paradox: A Model for the Role of Ku

7.1 Model for Ku in V(D)J Recombination and DSBR

To explain the DNA intermediates in terms of the biochemical properties of the Ku protein, we propose a model for the role of Ku in V(D)J recombination (see Fig. 4). In this model, the reaction proceeds in the following steps.

1. Concerted action of RAG1 and RAG2 introduces a nick 5' to the signal sequence, allowing Ku to bind to the nicked DNA intermediate. RAG1 and RAG2 then complete the cleavage reaction to form a hairpin coding end and a blunt signal end.

2. Upon creation of the hairpin end, Ku binds immediately because it has been preloaded onto the nicked DNA intermediate.

3. Activated DNA-PK is formed by the recruitment of DNA-PK$_{CS}$ to form a complex with Ku bound to the DNA. DNA-PK$_{CS}$ may serve an alignment function, so that two molecules of DNA-PK$_{CS}$ interact with each other to bring the appropriate coding ends in proximity to each other.

4. DNA-PK autophosphorylates its Ku subunit, conferring to Ku an ATP-dependent helicase activity. (Steps 3 and 4 may precede step 2, since they can occur once Ku is loaded onto the nicked DNA intermediate.)

5. Ku then unwinds the hairpin while moving inward from the end, exposing the hairpin and allowing the hairpin endonuclease to nick and open the hairpin.

6. Ku helicase unwinds the free DNA ends to allow the denatured single strands to align and pair in regions of microhomology.

7. In cases where pairing involves internal sequences, a DNA flap would be created from DNA distal to the region of homology. In fact, a flap endonuclease (FEN-1) has been characterized and purified (HARRINGTON and LIEBER 1994, 1995). FEN-1 specifically recognizes 5′ flap structures and cuts precisely at the base of the flap to leave a ligatable nick.

8. Alternatively, the DNA ends are aligned end to end and joined by either single-strand ligation followed by gap filling or blunt ligation. Thus, the joints can be made with or without microhomology. (In steps 6, 7, and 8, the opened hairpin is rejoined in a reaction identical to that used for general DSBR, as shown in Fig. 5.)

9. The two signal ends are brought together to form a signal joint in a reaction that requires Ku and the XR-1 gene product but not DNA-PK$_{CS}$. The rate of signal joint formation is likely to be much slower than for coding joints, since signal ends but not coding ends are readily detectable in wild-type thymocytes. Delayed formation of signal joints may be a mechanism for suppressing the unproductive fusion of a coding end to a signal end. Such hybrid joints are less likely if coding joints are already formed before signal joining begins.

7.2 Resolving the Scid Paradox

The scid paradox is resolved by the proposed model for Ku. In the model, the hairpin endonuclease gains access to transfected hairpins but not to V(D)J hairpins. In the case of transfected DNA, the hairpin endonuclease would nick the hairpin before Ku has a chance to bind, explaining the proper processing of transfected hairpins in scid cells. In the case of V(D)J recombination, the two step cleavage reaction by RAG1 and RAG2 first creates a nick adjacent to the signal sequence, allowing Ku to be loaded onto the DNA. When the hairpin is created, Ku is already present, thus denying access to the hairpin endonuclease. In scid cells deficient for DNA-PK$_{CS}$, Ku would fail to acquire

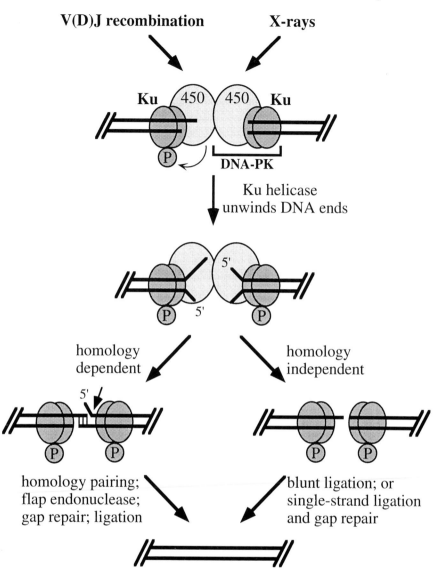

Fig. 5. Model for Ku in DSBR. Free DNA ends are created by X-rays or as intermediates in V(D)J recombination. DNA-PK$_{cs}$ has an alignment function and autophosphorylates Ku to give Ku helicase activity. Ku helicase unwinds the DNA ends, which can then be joined by homology dependent or homology independent mechanisms (see Fig. 3). The overall joining reaction is non-conservative and may involve the loss of nucleotides

Acknowledgments. The author thanks Rachel Gerstein, David Schatz, Kim Rathmell, and Vaughn Smider for helpful discussions. Supported in part by a grant from the Department of the Army, USA (DAMD17-94-J-4350).

References

Anderson C (1993) DNA damage and the DNA-activated protein kinase. Trends Biochem Sci 18:433–437

Beall E, Admon A, Rio D (1994) A *Drosophila* protein homologous to the human p70 Ku autoimmune antigen interacts with the P transposable element inverted repeats. Proc Natl Acad Sci USA 91:12681–12685

Biedermann K, Sun J, Giaccia A, Tosto L, Brown J (1991) Scid mutation in mice confers hypersensitivity to ionizing radiation and a deficiency in DNA double-strand break repair. Proc Natl Acad Sci USA 88:1394–1397

Blier PR, Griffith AJ, Craft J, Hardin JA (1992) Binding of Ku protein to DNA: measurement of affinity for ends and demonstration of binding to nicks. J Biol Chem 268:7594–7601

Blunt T, Finnie N, Taccioli G, Smith G, Demengeot J, Gottlieb T, Mizuta R, Varghese A, Alt F, Jeggo P, Jackson S (1995) Defective DNA-dependent protein kinase activity is linked to V(D)J recombination and DNA repair defects associated with the murine scid mutation. Cell 80:813–823

Bosma M, Carroll A (1991) The SCID mouse mutant: definition, characterization, and potential uses. Annu Rev Immunol 9:323–350

Boubnov N, Wills Z, Weaver D (1993) V(D)J recombination coding junction formation without DNA homology: processing of coding termini. Mol Cell Biol 13:6957–6968

Boubnov NV, Hall KT, Wills Z, Lee SE, He DM, Benjamin DM, Pulaski CR, Band H, Reeves W, Hendrickson EA, Weaver DT (1995) Complementation of the ionizing radiation sensitivity, DNA end binding, and V(D)J recombination defects of double-strand break repair mutants by the p86 Ku autoantigen. Proc Natl Acad Sci USA 92:890–894

Cai Q-Q, Plet A, Imbert J, Lafage-Pochitaloff M, Cerdan C, Blanchard J-M (1994) Chromosomal location and expression of the genes coding for Ku p70 and p80 in human cell lines and normal tissues. Cytogenet Cell Genet 65:221–227

Cao Q, Pitt S, Leszyk J, Baril E (1994) DNA-dependent ATPase from HeLa cells is related to human Ku autoantigen. Biochemistry 33:8548–8557

Chen DJ, Marrone BL, Nguyen T, Stackhouse M, Zhao Y, Siciliano MJ (1994) Regional assignment of a human DNA repair gene (XRCC5) to 2q35 by X-ray hybrid mapping. Genomics 21:423–427

Errami A, Smider V, Rathmell W, He D, Hendrickson E, Zdzienicka M, Chu G (1996) Ku86 defines the genetic defect and restores X-ray resistance and V(D)J recombination to complementation group 5 hamster cells. Mol Cell Biol 16:1519–1526

Falzon M, Fewell JW, Kuff EL (1993) EBP-80, a transcription factor closely resembling the human autoantigen Ku, recognizes single- to double-strand transitions in DNA. J Biol Chem 268:10546–10552

Feldmann H, Winnacker E (1993) A putative homologue of the human autoantigen Ku from *Saccharomyces cerevisiae*. J Biol Chem 268:12895–12900

Fulop G, Phillips R (1990) The scid mutation in mice causes a general defect in DNA repair. Nature 347:479–482

Gerstein R, Lieber MR (1993) Extent to which homology can constrain coding exon junctional diversity in V(D)J recombination. Nature 363:625–627

Getts R, Stamato T (1994) Absence of a Ku-like DNA end binding activity in the xrs double-strand DNA repair-deficient mutant. J Biol Chem 269:15981–15984

Gilfillan S, Dierich A, Lemeur M, Benoist C, Mathis D (1993) Mice lacking TdT: mature animals with an immature lymphocyte repertoire. Science 261:1175–1178

Gottlieb T, Jackson S (1993) The DNA-dependent protein kinase: requirement for DNA ends and association with Ku antigen. Cell 72:131–142

Hafezparast M, Kaur G, Zdzienicka M, Athwal R, Lehmann A, Jeggo P (1993) Subchromosomal localization of a gene (XRCC5) involved in double-strand break repair to the region 2q34-36. Somat Cell Mol Genet 19:413–421

Harrington JJ, Lieber MR (1994) The characterization of a mammalian DNA structure-specific endonuclease. EMBO J 13:1235–1246

Harrington JJ, Lieber MR (1995) DNA structural elements required for FEN-1 binding. J Biol Chem 270:4503–4508

Hendrickson E, Qin XQ, Bump E, Schatz D, Oettinger M, Weaver D (1991) A link between double-strand break related repair and V(D)J recombination: the scid mutation. Proc Natl Acad Sci USA 88:4061–4065

Jacoby D, Wensink P (1994) Yolk protein factor 1 is a *Drosophila* homolog of Ku, the DNA-binding subunit of a DNA-dependent protein kinase from humans. J Biol Chem 269:11484–11491

Jeggo AP (1990) Studies on mammalian mutants defective in rejoining double-strand breaks in DNA. Mutat Res 239:1–16

Kirschgessner C, Patil C, Evans J, Cuomo C, Fried L, Carter T, Oettinger M, Brown JM (1995) DNA-dependent kinase (p350) as a candidate gene for the murine SCID defect. Science 267:1178–1185

Komori T, Okada A, Stewart V, Alt F (1993) Lack of N regions in antigen receptor variable region genes of TdT-deficient lymphocytes. Science 261:1171–1175

Lafaille J, DeCloux A, Bonneville M, Takagaki Y, Tonegawa S (1989) Junctional sequences of T cell receptor γδ genes: implications for γδ T cell lineages and for a novel intermediate of V-(D)-J joining. Cell 59:859–870

Lee SE, Pulaski C, He DM, Benjamin DM, Voss JM, Um JY, Hendrickson E (1995) Isolation of mammalian mutants that are X-ray sensitive, impaired in DNA double-strand break repair and defective for V(D)J recombination. Mutation Res 336:279–291

Lewis S (1994a) The mechanism of V(D)J joining: lessons from molecular, immunological, and comparative analyses. Adv Immunol 56:27–150

Lewis SM (1994b) P nucleotide insertions, and the resolution of hairpin DNA structures in mammalian cells. Proc Natl Acad Sci USA 91:1332–1336

Li Z, Otevrel T, Gao Y, Cheng HL, Seed B, Stamato TD, Taccioli GE, Alt FW (1995) The XRCC4 gene encodes a novel protein involved in DNA double-strand break repair and V(D)J recombination. Cell 83:1079–1089

Lieber MR (1991) Site-specific recombination in the immune system. FASEB J 5:2934–2944

Lieber MR, Hesse JE, Lewis S, Bosma GC, Rosenberg N, Mizuuchi K, Bosma MJ, Gellert M (1988) The defect in murine severe combined immune deficiency: joining of signal sequences but not coding segments in V(D)J recombination. Cell 55:7–16

McBlane F, van Gent D, Ramsden D, Romeo C, Cuomo C, Gellert M, Oettinger M (1995) Cleavage at a V(D)J recombination signal requires only RAG1 and RAG2 proteins and occurs in two steps. Cell (in press)

McCormack W, Tjoelker L, Carlson L, Petryniak B, Barth C, Humphries E, Thompson C (1989) Chicken IgL gene rearrangement involves deletion of a circular episome and addition of single nonrandom nucleotides to both coding segments. Cell 56:785–791

Mimori T, Hardin JA (1986) Mechanism of interaction between Ku protein and DNA. J Biol Chem 261:10375–10379

Mimori T, Hardin JA, Steitz JA (1986) Characterization of the DNA-binding protein antigen Ku recognized by autoantibodies from patients with rheumatic disorders. J Biol Chem 261:2274–2278

Mimori T, Akizuki M, Yamagata H, Inada S, Yoshida S, Homma M (1981) Characterization of a high molecular weight acidic nuclear protein recognised by antibodies in sera from patients with polymyositis-scleroderma overlap. J Clin Invest 68:611–620

Mimori T, Ohosone Y, Hama N, Akizuki M, Homma M, Griffith AJ, Hardin JA (1990) Isolation and characterization of cDNA encoding the 80-kDa subunit protein of the human autoantigen Ku (p70/p80) recognized by autoantibodies from patients with scleroderma-polymyositis overlap syndrome. Proc Natl Acad Sci USA 87:1777–1781

Oettinger M, Schatz D, Gorka C, Baltimore D (1990) RAG-1 and RAG-2, adjacent genes that synergistically activate V(D)J recombination. Science 248:1517–1523

Paillard S, Strauss F (1991) Analysis of the mechanism of interaction of simian Ku protein with DNA. Nucl Acids Res 19:5619–5624

Paillard S, Strauss F (1993) Site-specific proteolytic cleavage of Ku protein bound to DNA. Proteins 15:330–337

Pergola F, Zdzienicka MZ, Lieber MR (1993) V(D)J recombination in mammalian cell mutants defective in DNA double-strand break repair. Mol Cell Biol 13:3464–3471

Rathmell WK, Chu G (1994a) A DNA end-binding factor involved in double-strand break repair and V(D)J recombination. Mol Cell Biol 14:4741–4748

Rathmell WK, Chu G (1994b) Involvement of the Ku autoantigen in the cellular response to DNA double-strand breaks. Proc Natl Acad Sci USA 91:7623–7627

Reeves W (1992) Antibodies to the p70/p80 (Ku) antigens in systemic lupus erythematosus. Rheum Dis Clin N Am 18:391–415

Reeves WH, Sthoeger ZM (1989) Molecular cloning of cDNA encoding the p70 (Ku) lupus auto-antigen. J Biol Chem 264:5047–5052

Roth DB, Wilson JH (1986) Nonhomologous recombination in mammalian cells: role for short sequence homologies in the joining reaction. Mol Cell Biol 6:4295–4304

Roth DB, Porter TN, Wilson JH (1985) Mechanisms of nonhomologous recombination in mammalian cells. Mol Cell Biol 5:2599–2607

Roth DB, Chang X, Wilson JH (1989) Comparison of filler DNA at immune, nonimmune, and on-cogenic rearrangements suggests multiple mechanisms of formation. Mol Cell Biol 9:3049–3057

Roth D, Menetski J, Nakajima P, Bosma M, Gellert M (1992) V(D)J recombination: broken DNA molecules with covalently sealed (hairpin) coding ends in scid mouse thymocytes. Cell 70:983–991

Schatz DG, Oettinger MA, Baltimore D (1989) The V(D)J recombination activating gene, RAG-1. Cell 59:1035–1048

Smider V, Rathmell WK, Lieber M, Chu G (1994) Restoration of X-ray resistance and V(D)J recom-bination in mutant cells by Ku cDNA. Science 266:288–291

Taccioli G, Rathbun G, Oltz E, Stamato T, Jeggo P, Alt F (1993) Impairment of V(D)J recombination in double-strand break repair mutants. Science 260:207–210

Taccioli G, Gottlieb T, Blunt T, Priestly A, Demengeot J, Mizuta R, Lehmann A, Alt F, Jackson S, Jeggo P (1994) Ku80: product of the XRCC5 gene and its role in DNA repair and V(D)J recom-bination. Science 265:1442–1445

Tuteja N, Tuteja R, Ochem A, Taneja P, Huang N, Simoncsits A, Susic S, Rahman K, Marusic L, Chen J, Zhang J, Wang S, Pongor S, Falaschi A (1994) Human DNA helicase II: a novel DNA unwinding enzyme identified as the Ku autoantigen. EMBO J 13:4991–5001

van Gent D, McBlane JF, Ramsden D, Sadofsky M, Hesse J, Gellert M (1995) Initiation of V(D)J recombination in a cell-free system. Cell 81:925–934

Yaneva M, Wen J, Ayala A, Cook R (1989) cDNA-derived amino acid sequence of the 86-kDa subunit of the Ku antigen. J Biol Chem 264:13407–13411

Characterization of Chinese Hamster Cell Lines That Are X-Ray-Sensitive, Impaired in DNA Double-Strand Break Repair and Defective for V(D)J Recombination

Sang Eun Lee, Dong Ming He, and Eric A. Hendrickson

1 Introduction

DNA double-strand breaks (DSB) can form spontaneously or be induced ex-
perimentally via ionizing radiation (IR). DSB are extremely toxic and all living
organisms have developed repair mechanisms to cope with these potentially
lethal lesions (Friedberg et al. 1995). In mammals, DNA DSB can also be introduced during the site-specific V(D)J recombination process, which is necessary for the assembly of immunoglobulin and T-cell receptor genes in B- and
T-cells, respectively. Elegant analyses of V(D)J recombination products in vivo
and in vitro strongly suggest that enzymatically induced DNA DSB are essential
intermediates in the V(D)J reaction process (Roth et al. 1992a,b; Schlissel et
al. 1993; van Gent et al. 1995). Thus, some aspects of DNA DSB repair and
mammalian V(D)J recombination may be related; the mechanism(s) underlying
these two processes is, however, still unknown.

   Historically, the isolation and characterization of mutants defective for a
particular function have repeatedly proven critical for elucidating the biochemical and molecular events involved in that function. The recent characterization
of mammalian IR-sensitive (IR^S) mutants has already provided considerable
information for our understanding of DNA DSB repair. Analysis of the murine
severe combined immune deficiency (*scid*) mutant (Biedermann et al. 1991;
Fulop and Phillips 1990; Hendrickson et al. 1991) and three Chinese hamster

Department of Molecular Biology, Cellular Biology and Biochemistry, Brown University, Providence,
RI 02912, USA

IR^S mutants (XR-1, *xrs*, and V3) (Pergola et al. 1993; Taccioli et al. 1993, 1994a) showed that these cell lines had defects in both DNA DSB repair and V(D)J recombination, confirming that the intermediates of DNA DSB repair and V(D)J recombination are indeed resolved by a common set of enzymes.

Recently, several laboratories have shown that the catalytic subunit of DNA-dependent protein kinase (DNA-PK) may be a candidate for the *scid* gene (Blunt et al. 1995; Kirchgessner et al. 1995; Lees-Miller et al. 1995). DNA-PK is a serine-threonine kinase consisting of at least two components: the 440 kDa catalytic subunit (DNA-PK$_{cs}$) and Ku protein (Dvir et al. 1992; Gottlieb and Jackson 1993; Suwa et al. 1994). Ku, a heterodimer of 70 and 86 kDa subunits, is a unique DNA binding protein that binds predominately, if not exclusively, to the ends of dsDNA (de Vries et al. 1989; Falzon et al. 1993; Mimori and Hardin 1986; Paillard and Strauss 1991). Ku is thus thought to provide the DNA binding component for the DNA-PK holoenzyme. Recently, it was shown that members of the fifth X-ray cross complementation group (XRCC5) (Thompson and Jeggo 1995; Zdzienicka 1995) lack Ku DNA end-binding activity (Getts and Stamato 1994; Rathmell and Chu 1994a,b). Since the Ku86 gene maps to human chromosome 2q33-35 (Cai et al. 1994), and XRCC5 group cells could be rescued by the same region (Jeggo et al. 1992), p86 Ku was a strong candidate for the *XRCC5* gene. Consistent with this hypothesis, several groups independently reported that transfection of a human Ku86 cDNA was able to partially rescue the defects of XRCC5 mutants (Boubnov et al. 1995; Smider et al. 1994; Taccioli et al. 1994b). Thus, DNA-PK has been unequivocally identified as an important mammalian DNA repair complex and mutations in either the DNA-PK$_{cs}$ or in the 86 kDa subunit of Ku result in severe X-ray sensitivity and DNA DSB repair and V(D)J recombinational defects.

Since analysis of cell mutants that are defective in DSB repair has proven to be a powerful technique, the availability of additional cell mutants which are defective in these processes can only help to unravel the mechanism of DSB repair. Recently, we isolated four IR^S mutants from a Chinese hamster lung V79-4 cell line (Lee et al. 1995). This report describes some of the progress made in our laboratory to characterize the molecular and biochemical defects in these mutants, which are sensitive to *X*-irradiation (*sxi*-1 to 4). First, we show that the hamster Ku86 cDNA functionally complements all the known defects of *sxi*-3 cells, suggesting that *sxi*-3 belongs to the XRCC5 complementation group. Secondly, we report the cloning of the hamster Ku70 gene. Lastly, we report that the current evidence demonstrates that the *sxi*-1 mutant is not defective in Ku70 as originally postulated, but has properties which may be attributable to a complex mutation.

2 Results and Discussion

2.1 Isolation of IR^S, DNA DSB-Repair Impaired and V(D)J Recombination Defective Mutants, *sxi*-1 to -4

After a brute force screen of 10 000 retrovirally infected cells using a replica-plating technique (JEGGO and KEMP 1983), we isolated four clones that were sensitive to X-irradiation (*sxi*-1 to -4; LEE et al. 1995). All of the *sxi* cell lines were also sensitive to the radiomimetic chemotherapeutic compound bleomycin and extremely sensitive to the radiomimetic topoisomerase inhibitor, etoposide (Table 1). Etoposide, a potent inhibitor of DNA topoisomerase II, stabilizes the enzyme in its DNA-bound state after introducing a DNA DSB (reviewed in LIU 1989). Thus, chromosomal DNA DSB associated with covalently bound protein appear to be particularly refractory to repair.

A subset of mammalian IR^S mutants have been shown to be defective in DNA DSB repair (reviewed in ZDZIENICKA 1995). A pulse-field gel electrophoresis assay was used to demonstrate that all of the *sxi* mutants were severely defective in the repair of DNA DSB (Table 1; LEE et al. 1995). Since all of the DSB repair defective mutants known to date had also been shown to be defective in V(D)J recombination, we next assessed the ability of *sxi* cells to perform either V(D)J coding or signal junction formation using an extrachromosomal V(D)J recombination assay (HESSE et al. 1987). All four mutant cell lines were deficient in both coding and signal junction formation (Table 1; BOUBNOV et al. 1995; LEE et al. 1995), distinguishing them from the *scid* mutation which affects only coding junction formation (HENDRICKSON et al. 1988; LIEBER et al. 1988) and extending the observation that mutants defective in DNA DSB repair are also defective in V(D)J recombination. Lastly, somatic cell hybrid analysis showed that *sxi*-2 and *sxi*-3 did not complement each other, or *xrs*-6 and, therefore, represent additional isolates of the XRCC5 complementation group (BOUBNOV et al. 1995). Moreover, several groups have reported that XRCC5 group cells are deficient in Ku DNA end-binding activity (GETTS and STAMATO 1994; RATHMELL and CHU 1994a,b). To extend this observation to our mutants, we examined DNA end-binding activity in the *sxi* mutants. Surprisingly, *sxi*-1 and *sxi*-4, in addition to *sxi*-2 and *sxi*-3, lacked DNA end-binding activity judging from DNA mobility shift assays (Table 1).

2.2 *sxi*-3 Cells Can Be Functionally Complemented by Ku86

Since *sxi*-3 cells belonged to the XRCC5 complementation group, it was anticipated that their defects could be rescued with a functional Ku86 gene. Indeed, stable transfection of a human Ku86 cDNA expression construct into *sxi*-3 cells could partially restore ionizing radiation resistance (IR^r), Ku DNA end-binding and V(D)J recombination activity (BOUBNOV et al. 1995). To extend these observations we recently cloned and sequenced a hamster Ku86 cDNA (HE et al. 1996). This gene was subcloned in both orientations into a eukaryotic expression vector and used to generate stable transfectants expressing either sense or anti-sense gene products. Expression of the sense orientation of this gene completely restored IR^r (Fig. 1b) and significantly restored Ku DNA end-binding (Fig. 2b) whereas cells expressing the anti-sense construct were as deficient in both activities as the sxi-3 cells themselves. In addition, we have shown that the hamster Ku86 gene will completely rescue the etoposide sensitivity and V(D)J proficiency of sxi-3 cells (Table 1; HE et al. 1996). Lastly, Northern analysis using the hamster cDNA as a probe demonstrated that sxi-3 cells are completely defective for Ku86 mRNA expression and that this expression was restored in cell lines stably expressing the sense orientation of the hamster Ku86 cDNA (Fig. 3b). From these experiments we conclude that the phenotypes of sxi-3 are directly related to the loss of endogenous Ku86 gene expression, and they confirm that Ku86 is an important component of the mammalian DNA DSB repair machinery.

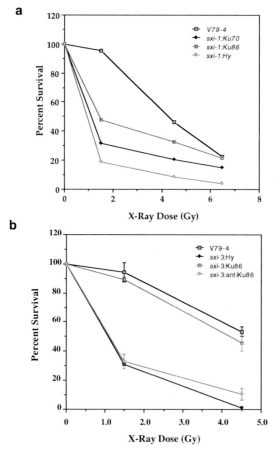

Fig. 1a,b. Functional complementation of the IR^S of sxi-1 and sxi-3. **a** V79-4, the parental, wild-type cell line; sxi-1:Ku70, sxi-1 cell line stably transfected with the Ku70 expression construct; sxi-1:Ku86, sxi-1 cell line stably transfected with the Ku86 expression construct; sxi-1:Hy, sxi-1 cell line stably transfected only with a hygromycin drug resistance marker. **b** V79-4, the parental, wild-type cell line; sxi-3:Hy, sxi-3 cell line stably transfected only with a hygromycin drug resistance marker; sxi-3:Ku86, sxi-3 cell line stably transfected with the Ku86 expression construct; sxi-3:antiKu86, sxi-3 cell line stably transfected with the Ku86 expression construct in the anti-sense orientation. Single cell suspensions were plated and then X-irradiated at the indicated doses. One week post-irradiation, cells were fixed and stained, and cells that survived to form colonies (50 cells) were scored

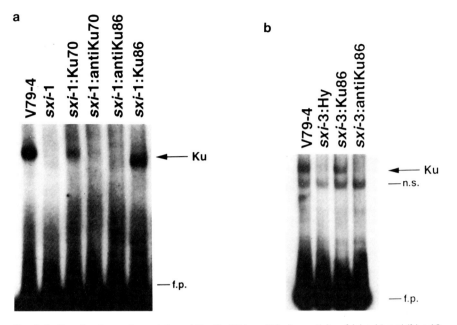

Fig. 2a,b. Functional complementation of the Ku DNA end-binding activity of (**a**) *sxi*-1 and (**b**) *sxi*-3 cells. Cell line descriptions are as given in the legend to Fig. 1. Nuclear extracts were prepared from the indicated cell lines, mixed with a radiolabelled ~150 bp dsDNA probe in the presence of 1 μg circular dsDNA and then analyzed using standard electrophoretic mobility shift assay conditions. *Ku*, the position of the authentic Ku:DNA complex; *n.s.*, a nonspecific complex occasionally observed; *f.p.*, free probe

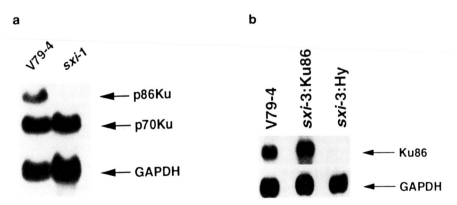

Fig. 3a,b. Northern analysis of **a** *sxi*-1 and **b** *sxi*-3 cells. PolyA⁺ mRNA was purified and prepared for Northern blot analysis. The corresponding filters were probed with the hamster Ku86 gene or a control glyceraldehye phosphate dehydrogenase (*GAPDH*) gene. In the case of *sxi*-1, the filter was also probed with the hamster Ku70. The positions of the three genes are shown with *arrows*

Table 1. Phenotypes of the sxi-1, sxi-2, sxi-3 and sxi-4 cell lines

Cell Line	IR[a]	Bleomycin[b]	Etoposide[c]	DSB Repair[d]	V(D)J[e]	Ku DEB[f]
V79-4	R	R	R	+	+	+
sxi-1	S	S	S	–	–	–
sxi-2	S	S	S	–	–	–
sxi-3	S	S	S	–	–	–
sxi-4	S	S	S	–	–	–
sxi-1:Ku86	R/S	ND	ND	ND	±	±
sxi-3:Ku86	R	ND	R	ND	+	+

R, resistant; S, sensitive; R/S, partially resistant; ND, not done; +, proficient; -, deficient; ±, partially
proficient
[a] Ionizing radiation sensitivity
[b] Bleomycin sensitivity
[c] Etoposide sensitivity
[d] The ability to repair DNA DSB as assessed by PFGE analysis
[e] The ability to carry out either V(D)J recombination coding or signal junction formation as assessed
by extrachromosomal transfection assays
[f] An activity that binds to the ends of a dsDNA fragment in the presence of excess circular dsDNA
as assessed by electrophoretic mobility shift assays

2.3 sxi-1 Is Not Defective in Ku70 and May Represent a Complex Mutation

Since sxi-1 was also defective in Ku DNA end-binding activity, we considered
the possibility that it might be defective in Ku70. To pursue this hypothesis,
we cloned a Ku70 cDNA from a Chinese hamster V79 cell library by its ability
to cross-hybridize with the human Ku70 cDNA probe. Hamster and human
Ku70 (GRIFFITH et al. 1992; REEVES and STHOEGER 1989) are 84% identical (99
amino acid changes) and hamster and mouse Ku70 (PORGES et al. 1990) are
92% identical with only 49 amino acid changes (Fig. 4). The hamster Ku70
contains two conserved, but imperfect, leucine heptad repeats which may
function as a leucine zipper for protein-protein interaction with Ku86 (MIMORI
et al. 1990). Moreover, hamster Ku70 has a conserved C-terminal helix-turn-helix
motif and a conserved N-terminal acidic amino acid domain which might serve
as DNA binding and activation domains, respectively (Fig. 4). We utilized the
hamster Ku70 DNA sequence to generate oligonucleotide primers for reverse
transcription polymerase chain reaction (RT-PCR) and used these to clone the
sxi-1 Ku70 allele. Sequence analysis of sxi-1 Ku70 alleles demonstrated that
they were completely wild-type (data not shown), ruling out a mutation in
Ku70 as the cause of the phenotype in sxi-1. To gain insight into the molecular
basis of the sxi-1 defect, we next established stable transfectants of sxi-1
cells carrying either the sense or anti-sense orientations of either the hamster
Ku70 or Ku86 cDNAs. Hamster Ku70 restored a small, but reproducible, amount
of IR[r] to sxi-1 cells (Fig. 1a) The hamster Ku86 gene complemented sxi-1 cells
better than Ku70, but in contrast to sxi-3 cells, the IR[r] complementation was
only partial (compare Fig. 1a with 1b). Similarly, stable transfectants expressing

```
Hamster     1   MSGWESYYKT EGEEE*EEE* EESPDPGGEY KYSGRDSLIF LVDASRAMFD
Murine          ?????????? ???--*---* -----T---- ---------- ---------E
Human           ---------- --D--A---Q --NLEAS-D- ---------- -----K---E

Hamster    51   SQGEDEITPF DMSIQCIQSV YTSKIISSNR DLLGVVFYGT EKDKNSVNFK
Murine          ------L--- ---------- --------D- ---A------ ----------
Human           --S---L--- ---------- --------D- ---A------ ----------

Hamster   101   NIYVLQELDN PGAKRVLELD QFKGQQGKKH FQDTIGHGSD YSLSEVLWVC
Murine          ------D--- ---------- ---------- -R--V----- ----------
Human           ---------- -----I---- -------Q-R ---MM----- ----------

Hamster   151   ANLFSDVQVK MSHKRIMLFT NEDDPHGNDS AKASRARTKA NDLRDTGIFL
Murine          --------L- ---------- -------R-- ---------- S---------
Human           --------F- ---------- ---N------ ---------- G---------

Hamster   201   DLMHLKRRGG FDISLFYRDI MSIAEDEDLG VHFEESSKLE DLLRKVRAKE
Murine          ------KP-- --V-V----- ITT------- ---------- ----------
Human           ------KP-- ---------- I--------R ---------- ----------

Hamster   251   TKKRVLSRLR FKLGKDVALM VGIYNLIQKA NKPFPVRLYR ETNEPVKTKT
Murine          ---------K ----E--V-- ------V--- ---------- ----------
Human           -R--A----K L--N--IVIS ------V--- L--P-IK--- ----------

Hamster   301   RTFNVNTGSL LLPSDTKRSQ TYGSRQIVLE KEETEELKRF DEPGLILMGF
Murine          ---------- --------L ---T------ ---------- ----------
Human           ----TS--G- ---------- I------I-- ---------- -D---MLMGF

Hamster   351   KALVMLKKHH YLRPSLFVYP EESLVNGSST LFSALLTKCL EKEVMAVCRY
Murine          -PT-----Q- ---------- -----S---- ---------V --K-I-----
Human           -P--L----- ---------- -----I---- ------I--- ----A-L---

Hamster   401   TSRKNVPPYF VALVPQEEEL DDQNIQVTPA GFQLVFLPYA DDKRKVPFTE
Murine          -H----S--- ---------- --------G ---------- ----------
Human           -P-R-I---- ---------- ---K-----P --------F- -----M----

Hamster   451   KVMANPEQID KMKAIVHNVR FTYRSDSFEN PVLQQHFRNL EALALDMMES
Murine          --T--Q---- ------Q--- ---------- ---------- ----------
Human           -I--T---VG ------EKL- ---------- ---------- ------L--P

Hamster   501   EQVVDLTLPK AEAIKKRLGS LADEFKELVY PPGYNPEGKA TKRKQDDEGS
Murine          ---------- V--------- ---------- --------V A---------
Human           --A------- V--MN----- -V-------- --D------V ----H-N---

Hamster   551   ASKKPKEELS EEELKAHFAK GTLGKLTVPT LKEVCKAYGL KSGPKKQELL
Murine          T-----V--- --------R- ---------- --DI---H-- ----------
Human           G--R--V-Y- -----T-IS- -----F---M ---A-R---- ---L------

Hamster   601   DALTRHFQKN
Murine          ---I--LE--
Human           E---K---D*
```

Fig. 4. The predicted amino acid sequence of the hamster Ku70 gene. The murine (PORGES et al. 1990) and human (GRIFFITH et al. 1992; REEVES and STHOEGER 1989) sequences are shown for comparison. The published mouse sequence was from a partial cDNA and the N-terminal 13 amino acids (shown by *question marks*) have not been described. The two conserved, but imperfect, leucine heptad repeats at amino acids 256–276 and 366–386 are shown in bold. A Chinese hamster cDNA library was screened with the human Ku70 gene and a single 700 bp partial cDNA was obtained that corresponded to the middle of Ku70. This sequence was used to design PCR primers to isolate the N- and C-terminal coding sequences by 5'- and 3'-RACE reactions

Roth DB, Menetski JP, Nakajima PB, Bosma MJ, Gellert M (1992a) V(D)J recombination: broken DNA molecules with covalently sealed (hairpin) coding ends in scid mouse thymocytes. Cell 70:983–991

Roth DB, Nakajima PB, Menetski JP, Bosma MJ, Gellert M (1992b) V(D)J recombination in mouse thymocytes: double-strand breaks near T cell receptor δ rearrangement signals. Cell 69:41–53

Schlissel M, Constantinescu A, Morrow T, Baxter M, Peng A (1993) Double-strand signal sequence breaks in V(D)J recombination are blunt, 5'-phosphorylated, RAG-dependent, cell-cycle regulated. Genes Dev 7:2520–2532

Smider V, Rathmell WK, Lieber MR, Chu G (1994) Restoration of X-ray resistance and V(D)J recombination in mutant cells by Ku cDNA Science 266:288–291

Suwa A, Hirakata M, Takeda Y, Jesch SA, Mimori T, Hardin JA (1994) DNA-dependent protein kinase (Ku protein-p350 complex) assembles on double-stranded DNA. Proc Natl Acad Sci, USA 91:6904–6908

Taccioli GE, Rathbun G, Oltz E, Stamato T, Jeggo PA, Alt FW (1993) Impairment of V(D)J recombination in double-strand break repair mutants. Science 260:207–210

Taccioli GE, Cheng H-L, Varghese AJ, Whitmore G, Alt FW (1994a) A DNA repair defect in Chinese hamster ovary cells affects V(D)J recombination similary to the murine *scid* mutation. J Biol Chem 269:7439–7442

Taccioli GE, Gottlieb TM, Blunt T, Priestley A, Demengeot J, Mizuta R, Lehmann AR, Alt FW, Jackson SP, Jeggo PA (1994b) Ku80: Product of the XRCC5 gene and its role in DNA repair and V(D)J recombination. Science 265:1442–1445

Thompson LH, Jeggo PA (1995) Nomenclature of human genes involved in ionizing radiation sensitivity. Mutat Res (in press)

van Gent DC, McBlane JF, Ramsden DA, Sadofsky MJ, Hesse JE, Gellert M (1995) Initiation of V(D)J recombination in a cell-free system. Cell 81:925–934

Zdzienicka MZ (1995) Mammalian mutants defective in the response to ionizing radiation-induced DNA damage. Mutat Res 336:203–213

Identification of the *XRCC4* Gene: Complementation of the DSBR and V(D)J Recombination Defects of XR-1 Cells

Zhiying Li and Frederick W. Alt

1 Introduction

Antigen receptor variable region genes are assembled from component gene segments during the differentiation of B and T lymphocytes (Lansford et al. 1995). This process, termed V(D)J recombination, is initiated by RAG-1 and RAG-2, the only lymphoid-specific proteins required for this reaction (Schatz et al. 1989; Oettinger et al. 1990; Mombaerts et al. 1992; Shinkai et al. 1992). Expression of RAG-1 and RAG-2 promotes cleavage at the junction of the V, D, or J coding segments and their flanking recognition sequences (RS), generating hairpin-like coding ends and blunt RS ends (van Gent et al. 1995). Subsequent joining of the liberated DNA ends appears to involve more generally expressed proteins including several proteins that also participate in DNA double-strand break repair (DSBR) (reviewed by Taccioli and Alt 1995).

Elucidation of factors shared by DSBR and V(D)J recombination has emerged from studies of several radiosensitive Chinese hamster ovary (CHO) cell lines, including XR-1, xrs-6 and V-3 (reviewed by Taccioli and Alt 1995). These three cell lines represent X-ray repair cross complementing groups (XRCC)-4, 5 and 7, respectively, and their radiosensitivities are specifically caused by defective DSBR (reviewed by Jeggo et al. 1991 and Taccioli and Alt 1995). When RAG-1 and RAG-2 are ectopically expressed to provide V(D)J recombination ability, all three of these mutant hamster cell lines are capable of supporting initiation of the V(D)J recombination reaction. However, the xrs-6 and XR-1 lines are severely impaired in ability to join coding and RS ends

Center for Blood Research and Department of Genetics, Harvard Medical School, 200 Longwood Avenue, Boston, MA 02115, USA

(TACCIOLI et al. 1992, 1993), whereas V-3 cells, similar to cells homozygous for the murine *scid* mutation, are preferentially impaired in joining coding ends (TACCIOLI et al. 1992, 1994a; PERGOLA et al. 1993). Somatic cell hybrid studies confirmed that V-3 and *scid* reside in the same complementation group (TACCIOLI et al. 1994a).

The defects in xrs-6 and V-3 cells involve deficiencies in expression of two components of the DNA-dependent protein kinase (DNA-PK) complex. Ku80, which along with Ku70 forms the DNA-binding subunit of DNA-PK, is defective in xrs-6 cells and p460 (DNA-PKcs), which is the catalytic component of DNA-PK, is defective in V-3/Scid cells. Therefore, Ku80 represents the *XRCC5* gene and DNA-PKcs represents the *XRCC7* gene (reviewed by TACCIOLI and ALT 1995). XR-1, despite having similar defects to xrs-6 in V(D)J recombination, have normal DNA-PK and Ku DNA-end binding activities (GETTS and STAMATO 1994; RATHMELL and CHU 1994; BLUNT et al. 1995), and expression of human DNA-PK components failed to correct the XR-1 defects (TACCIOLI et al. 1994b; BLUNT et al. 1995). Furthermore, XR-1 has a unique cell-cycle-dependent radiosensitivity, being hypersensitive to X irradiation during the G1 and early S phases of the cell cycle, but showing nearly wild-type levels of resistance in late S phase (STAMATO et al. 1983). A putative human gene, termed *XRCC4*, which can complement the XR-1 defects, has been mapped to human chromosome 5q13-14 (GIACCIA et al. 1990; TACCIOLI et al. 1993; OTEVREL and STAMATO 1995).

2 Complementation Cloning of *XRCC4*

To further elucidate the XR-1 defect, we first characterized the ability of these cells to support inversional V(D)J recombination within the stably integrated G12 substrate (LI et al. 1995; Fig. 1). Normal recombination of the G12 construct inverts a *gpt* gene and activates its expression, conferring resistance to mycophenolic acid (MPA) (YANCOPOULOS et al. 1990). Since inversional recombination demands formation of both coding and RS joins and XR-1 cells are defective in both, we expected G12 inversion to occur at a very low frequency. Correspondingly, we found that RAG-expressing XR-1 cells initiated V(D)J recombination of the G12 construct at a substantial rate, but most of the rearrangement products were aberrant, involving deletion of the *gpt*-containing sequence (LI et al.1995). When pools of XR-1 RAG-transfectants were selected for growth in MPA, the frequency of MPA-resistant cells was 10^{-5} or less, at least 1000-fold lower than wild type cells (LI et al.1995). In contrast, murine Scid cells, which have a relatively unimpaired level of RS joining, show a much higher rate of inversion of a very similar construct (HENDRICKSON et al. 1990).

The very low frequency of inversional recombination in XR-1 cells confirmed the feasibility of isolating the *XRCC4* gene by a complementation strategy

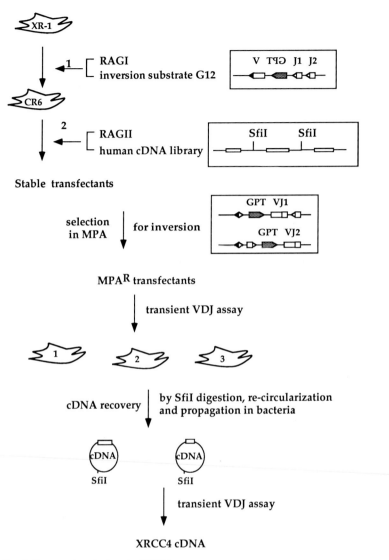

Fig. 1. Strategy for cloning the *XRCC4* gene

(outlined in Fig. 1; LI et al. 1995). For this purpose, we first established an XR-1 cell line, designated CR6, that contained one copy of the G12 construct and multiple active copies of RAG-1 expression vectors. Subsequently, RAG-2 expression vectors and DNA from a human teratocarcinoma cDNA library (in a shuttle vector that permitted both mammalian expression and bacterial propagation; BIERHUIZEN et al. 1993) were co-transfected into CR6. The transfected cDNA sequences were concatemerized at the unique SfiI site to promote uptake of multiple sequences and facilitate subsequent rescue. Potentially complemented XR-1 cells, presumably exhibiting G12 inversion, were selected based on growth in the presence of MPA. Actual complementation of the V(D)J recombination defect in the MPA-resistant cells was confirmed by assaying their ability to support transient V(D)J recombination. Three MPA-resistant clones displayed a significantly enhanced rate and fidelity of V(D)J recombination when assayed with pJH200, a substrate that measures RS join formation (HESSE et al. 1987; LI et al. 1995).

The cDNA clones that were integrated into these three cell lines were rescued by digestion of the genomic DNA with SfiI, recircularization, and propagation in host bacteria (Fig. 1; LI et al. 1995). Individual cDNA clones were analyzed for their ability to correct the XR-1 V(D)J recombination defects in transient assays. One of these clones was integrated into the genome of all three complemented XR-1 cell lines. This clone (termed *XRCC4*, see below) was found to completely restore V(D)J recombination ability of XR-1 cells when introduced either transiently or stably (LI et al. 1995; Table 1). Stable expression of the *XRCC4* cDNA sequence also significantly restored radioresistance to XR-1 cells (LI et al. 1995). The complementing effect, however, was limited to XR-1: transient expression of this sequence did not complement the V(D)J recombination defects of either xrs-6 or V-3 cells (LI et al. 1995). Furthermore, stable expression of XRCC4 did not correct either the V(D)J recombination or X-ray sensitivity defects of the V-3 cells (Fig. 2; Table 1).

Table 1. Analysis of signal and coding joining in CHO cells by Transient V(D)J recombination assay

| Cell line | cDNA Plasmid | pJH200 (Signal) | | | PJH290 (Coding) | | |
		AmpRCamR AmpR	%	Relative level	AmpRCamR AmpR	%	Relative level
4326A (wt)	–	28/347	8.1	1	220/4470	4.9	1
XR-1	–	118/19650	0.6	<0.1	ND		
XR-1	XRCC4	504/2640	19	2.3	365/6350	5.7	1.2
XR-1.XRCC4.1	–	348/3900	8.9	1.1	187/3950	4.7	0.9
AA8 (wt)	–	112/22500	0.5	1	48/1670	2.9	1
V-3	–	71/2220	3.2	6.4	8/15300	0.052	0.018
V-3	XRCC4	60/4350	1.4	2.8	3/20550	0.015	0.005
V-3.XRCC4.1	–	81/14400	0.6	1.1	0/38850	<0.003	<0.001
V-3.XRCC4.2	–	11/1635	0.7	1.3	6/32400	0.019	0.007
V-3.XRCC4.3	–	94/4830	1.9	3.9	7/37650	0.019	0.007
V-3.XRCC4.5	–	15/405	3.7	7.4	13/96000	0.014	0.005
V-3.XRCC4.6	–	46/5700	0.8	1.6	1/2145	0.047	0,016

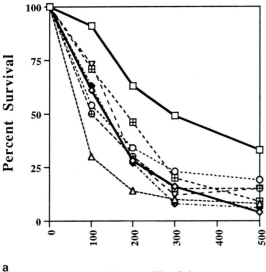

a

Dose (Rads)

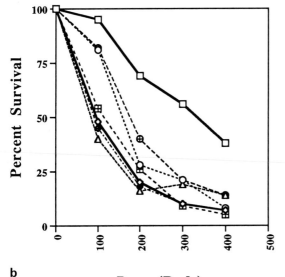

b

Dose (Rads)

Fig. 2a,b. X-ray irradiation survival curves for a series of V-3 transfectants. *AA8 (open squares),* parental wild type cell line for V-3 (*open rhombus*). V-3.XRCC4.1–6 (**a**) are various V-3 cell lines stably transfected with pPGK-Puro and the human *XRCC4* cDNA (*open circles,* V-3.XRCC4.1; *triangles,* V-3.XRCC4.2; *crossed squares,* V-3.XRCC4.3; *filled rhombus,* V-3.XRCC4.5; *crossed circles,* V-3.XRCC4.6). Expression of the human XRCC4 protein was confirmed by Western blotting analysis. V-3.1–8 (**b**) are various V-3 cell lines transfected with pGK-Puro only (*open circles,* V-3.1; *upward triangles,* V-3.3; *crossed squares,* V-3.5; *filled rhombus,* V-3.6; *crossed circles,* V-3.7; *downward triangles,* V-3.8). The purpose in including V-3.1–8 is to show the range of X-ray sensitivity among different subclones of V-3. X-ray irradiation tests were performed as follows: 100–200 cells were seeded in each well of a six-well plate. After incubation for overnight, the cells were irradiated with the gamma ray dosage indicated at a dose rate of 100 rad/min with a 662 Kev Cs^{147} irradiator. At 4–5 days after irradiation, colonies of cells were stained by crystal violet/EtOH solution and numbers of colonies counted. Percentage survival was calculated with the colony number of nonirradiated cells taken as a standard. Each data point is the mean of duplicate experiments

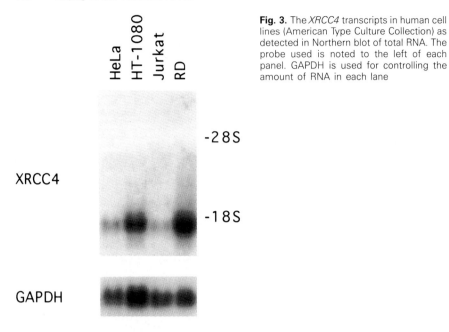

Fig. 3. The *XRCC4* transcripts in human cell lines (American Type Culture Collection) as detected in Northern blot of total RNA. The probe used is noted to the left of each panel. GAPDH is used for controlling the amount of RNA in each lane

3 Characterization of the *XRCC4* Gene

The long open reading frame of the XRCC4 cDNA can encode an approximately 40Kd protein that does not have significant homology to any previously described, although it does contain a potential nuclear localization motif and several potential phosphorylation sites (LI et al. 1995). We mapped the gene encoding the *XRCC4* cDNA sequence to human chromosome 5q11.2–13.3 (LI et al. 1995), showing that the sequence we isolated derived from the chromosomal region previously implicated (OTEVREL and STAMATO 1995). The mouse homologue of the gene was also isolated (LI et al. 1995) and mapped to a region in mouse chromosome 13 that is syntenic to human 5q13 (unpublished observation by Y. GAO, F. W. ALT, and N. JENKINS). No relevant mouse or human disease yet has been identified in these regions. The *XRCC4* gene encodes a ubiquitously expressed transcript of approximately 1.8kb in both human (Fig. 3) and mouse (LI et al. 1995). Southern blotting analyses revealed that the XR-1 mutation involved a complete deletion of the *XRCC4* gene (LI et al. 1995). Therefore, lack of XRCC4 expression is not cell lethal, and it may be possible to further analyze its function in vivo through a gene-targeted mutational approach.

4 Potential Functions of the *XRCC4* Protein

The XRCC4 sequence does not give any significant clues to its function. Any speculation would need to account for the requirement for the XRCC4 protein in the end-joining reactions that occur both during V(D)J recombination and DSBR. Presumably, XRCC4 plays the same role in both processes, although this remains to be determined. In particular, it is still unclear why expression of human XRCC4 completely restores the V(D)J recombination ability but only partially corrects X-ray sensitivity defect in XR-1 cells (Li et al. 1995).

XRCC4, like Ku80, could be involved in DNA end protection, although XR-1 cells do not have apparent DNA end-binding defects (Rathmell and Chu 1994; Getts and Stamato 1994). XRCC4 could also be involved in DNA damage recognition, possibly recruiting repair proteins to the site of the DNA break, analogous to the proposed functions of the XPB-XPD proteins in excision repair (Drapkin et al. 1994). The XRCC4 protein may also contain or regulate an enzymatic activity involved in ligation of DNA ends. Another DNA repair factor, XRCC1, has been shown to be associated with DNA ligase III (Caldecott et al. 1994). Alternatively, XRCC4 may have a role in maintaining the spatial structure of the putative recombination/repair complex, such that its absence leads to inability to efficiently align and ligate the appropiate ends. In this context, XR-1 cells have been shown to have abnormal contents of certain DNA-anchoring matrix proteins (Malyapa et al. 1994). The above proposed roles (and many other conceivable ones) for XRCC4 are not mutually exclusive and await further biochemical and genetic studies.

Acknowledgement. F. W. Alt is supported by the Howard Hughes Medical Institute and by NIH grants AI20047 and AI35714.

References

Bierhuizen MFA, Mattei MM, Fukuda M (1993) Expression of the development I antigen by a cloned human cDNA encoding a member of a beta-1,6-N-acetylglucosaminyltransferase gene family. Genes Dev 7:468–478

Blunt T, Finnie NJ, Taccioli GE, Smith GCM, Demengeot J, Gottlieb TM, Mizuta R, Varghese AJ, Alt FW, Jeggo PA, Jackson SP (1995) Defective DNA-dependent protein kinase activity is linked to V(D)J recombination and DNA repair defects associated with the murine *scid* mutation. Cell 80:813–823

Caldecott KW, McKeown CK, Tucker JD, Ljungquist S, Thompson LH (1994) An interaction between the mammalian DNA repair protein XRCC1 and DNA ligase III. Mol Cell Biol 14:68–76

Drapkin R, Sancar A, Reinberg D (1994) Where transcription meets repair. Cell 77:9–12

Getts RC, Stamato TD (1994) Absence of a Ku-like DNA end binding activity in the xrs double-strand DNA repair-deficient mutant. J Biol Chem 269(23):15981–15984

Giaccia AJ, Denko N, MacLaren R, Mirman D, Waldren C, Hart I, Stamato TD (1990) Human chromosome 5 complements the DNA double-strand break-repair deficiency and gamma-ray sensitivity of the XR-1 hamster variant. Am J Hum Genet 47:459–469

Hendrickson EA, Schlissel MS, and Weaver DT (1990) Wild-type V(D)J recombination in *scid* pre-B cells. Mol Cell Biol 10(10):5397–5407

Hesse JE, Lieber MR, Gellert M, Mizuuchi K (1987) Extrachromosomal DNA substrates in pre-B cells undergo inversion or deletion at immunoglobulin V-(D)-J joining signals. Cell 49:775–783

Jeggo PA, Tesmer J, Chen DJ (1991) Genetic analysis of ionising radiation sensitive mutants of cultured mammalian cell lines. Mut Res 254:125–133

Lansford R, Okada A, Chen J, Oltz EM, Blackwell TK, Alt FW, Rathbun G (1995) Mechanism and control of immunoglobulin gene rearrangement. In: Hames BD (ed) Molecular immunology, 2nd edn. IRL, Oxford (in press)

Li Z, Otevrel T, Gao Y, Cheng H, Seed B, Stamato TD, Taccioli GE, Alt FW (1995)The XRCC4 gene encodes a novel protein involved in DNA double-strand break repair and V(D)J recombination. (in press)

Malyapa RS, Wright WD, Roti JLR (1994) Radiation sensitivity correlates with changes in DNA supercoiling and nucleoid protein content in cells of three Chinese hamster cell lines. Radiat Res 140:312–320

Mombaerts P, Iacomini J, Johnson RS, Herrup K, Tonegawa S, Papaionnou VE (1992) RAG-1 deficient mice have no mature B and T lymphocytes. Cell 68:869–877

Oettinger MA, Schatz DG, Gorka C, Baltimore D (1990) RAG-1 and RAG-2, adjacent genes that synergistically activate V(D)J recombination. Science 24:1517–1522

Otevrel T, Stamato TD (1995) Regional localization of the XRCC4 human radiation repair gene. Genomics 27:211–214

Pergola F, Zdzienicka MZ, Lieber MR (1993) V(D)J recombination in mammalian cell mutants defective in DNA double-strand break repair. Mol Cell Biol 13:3464–3471

Rathmell WK, Chu G (1994) A DNA end-binding factor involved in double-strand break repair and V(D)J recombination. Mol Cell Biol 14:4741–4748

Schatz DG, Oettinger MA, Baltimore D (1989) The V(D)J recombination activating gene, RAG-1. Cell 59:1035–1048

Shinkai Y, Rathbun G, Lam K-P, Oltz EM, Stewart V, Mendelsohn M, Charron J, Datta M, Young F, Stall AM, Alt FW (1992) RAG-2-deficient mice lack mature lympocytes owing to inability to initiate V(D)J rearrangement. Cell 68:855–867

Stamato TD, Weinstein R, Giaccia A, Mackenzie L (1983) Isolation of cell cycle-dependent gamma-ray-sensitive Chinese hamster ovary cell. Somat Cell Genet 9:165–173

Taccioli GE, Alt FW (1995) Potential targets for autosomal SCID mutations. Curr Opin Immunol 7:436–440

Taccioli GE, Rathbun G, Shinkai Y, Oltz EM, Cheng H, Whitmore G, Stamato T, Jeggo P, Alt FW (1992) Activities involved in V(D)J recombination. Curr Topics Microbiol Immunol 182:107–114

Taccioli GE, Rathbun G, Oltz E, Stamato T, Jeggo PA, Alt FW (1993) Impairment of V(D)J recombination in double-strand break repair mutants. Science 260:207–210

Taccioli GE, Cheng H, Varghese AJ, Whitmore G, Alt FW (1994a) A DNA repair defect in Chinese hamster ovary cells affects V(D)J recombination similarly to the murine *scid* mutation. J Biol Chem 269:1–4

Taccioli GE, Gottlieb TM, Blunt T, Priestley A, Demengeot J, Mizuta R, Lehmann AR, Alt FW, Jackson SP, Jeggo PA (1994b) Ku80: product of the *XRCC5*gene and its role in DNA repair and V(D)J recombination. Science 265:1442–1445

van Gent DC, McBlane JF, Ramsden DA, Sadofsky MJ, Hesse JE, Gellert M (1995) V(D)J cleavage in a cell-free system. Cell 81:926–934

Yancopoulos GD, Nolan GP, Pollock R, Prockop S, Li SC, Herzenberg LA, Alt FW (1990) A novel fluorescence-based system for assaying separating live cells according to VDJ recombinase activity. Mol Cell Biol 10:1697–1704

Developmental and Molecular Regulation of Immunoglobulin Class Switch Recombination

Matthias Lorenz and Andreas Radbruch

1 Introduction

The ability of B lymphocytes to perform immunoglobulin (Ig) class switch recombination depends largely on their differentiation stage. Pro-B cells, pre-B cells, and plasma cells do not switch at detectable levels. B cell receptor (BCR) cross linking, T cell help, and cytokines can drive the mature B cells into proliferation and differentiation and induce switch recombination. This recombination is targeted to distinct switch regions by cytokines through the induction of switch transcripts (Lorenz et al. 1995a,b). Cytokines also control the frequency of detectable, switched cells by regulating the amount of Ig secreted and the clone size of switched cells (for review, see Vercelli and Geha 1992; Coffman et al. 1993; Harriman et al. 1993; Lorenz and Radbruch 1996). Here, we focus on those molecules, which are involved in the regulation of class switching.

The recent technology of introducing targeted mutations into the murine germline is enabling us to define the molecules involved in cellular differen-

Institute for Genetics, Weyertal 121, 50931 Cologne, Germany

tiation ad libitum. Already now, targeted mice have contributed considerably to the molecular understanding of Ig class switch recombination and its regulation. For a number of genes and DNA sequences it has been shown that they affect the frequency of switched B lymphocytes Some of them and their effects are listed in Table 1. It is obvious that these genes and their products act on a variety of levels, which are discussed in detail below. Reduced or enhanced Ig levels in the serum of homozygous targeted mice could also be due to an intrinsic defect in B or helper cell functions or to a defect in other cells which influence the production of Ig in general. Defects in BCR cross linking, antigen presentation, costimulatory signals, cytokine production or control elements for switch transcripts can dramatically influence the frequency of class switching, although many of them do not interfere with class switching as such. On the other hand, the importance of particular genes involved in switching will not become immediately evident, if the functions of these genes are backed up by redundant genes.

2 Developmental Regulation of Immunoglobulin Class Switching

It is still not entirely clear at which point of development mammalian B cells acquire the ability to perform Ig class switching. Class switching of pro- and pre-B cells in the bone marrow has never been reported unambiguously, but also not rigorously excluded for low frequencies. Pre-B cells isolated from the bone marrow express only cytoplasmic μ heavy chains (BURROWS et al. 1979), but no cytoplasmic γ or α heavy chains (KUBAGAWA et al. 1982). In an extrachromosomal recombination assay, switch substrates are predominantly recombined in B cell lines representing the mature B cell stage and less in pre-B cell lines or plasma cell lines (DANIELS and LIEBER 1995). In conceptional terms, class switching is not required for B cells until their first contact with antigen. Before the onset of somatic hypermutation, the primordial low affinity–high avidity IgM with its ten binding sites per molecule seems to be best suited to combat with antigen. The class-specific targeting of switch recombination, as discussed below, would make sense only in antigen-specific activations, and not in pro-B and pre-B cells, which lack an antigen receptor. The physiological relevance of the observation that in murine 18.81 Abelson-transformed pre-B cells, switch recombination from IgM to IgG2b occurs at low frequencies upon stimulation with lipopolysaccharide (LPS) (BURROWS et al. 1981, 1983) remains unclear.

3 The Onset of Switch Recombination

3.1 Differential Activation of B Cells by Antigens

It has been known for some time that particular antigens are correlated to particular Ig isotypes in the respective immune responses and that those immune responses are to varying degrees dependent on the help of T lymphocytes. Responses to most polysaccharides are largely independent of T cell help, responses to T independent type 1 (TI-1), or type 2 antigens (TI-2)) being dominated by IgG3. Responses to soluble protein antigens mostly depend on T cell help (TD antigens) and often are dominated by IgG1 (for review, see LORENZ and RADBRUCH 1996). TI-1, TI-2, and TD antigens use different pathways for B cell activation, i.e., the stimulation of murine resting B cells to enter the G1 phase of cell cycle. For TI-2 and TD antigens, the BCR and B cell coreceptor complex transmit the activation signal, for TI-1 antigens other B cell surface molecules like CD14 on human B cells are involved, acting as mitogen receptors (ZIEGLER-HEITBROCK et al. 1994). Although both being antigen-specific activations, TI-2 and TD responses are genetically dissected by the X-linked immunodeficiency (*Xid*), Bruton's tyrosine kinase (*btk*), major histocompatibility complex (MHC) class II, CD19 and CD40/CD40L mutations (Table 1). The TI-2, but not the TD immune response is abolished in Xid (KHAN et al. 1995) and btk (KERNER et al. 1995; KHAN et al. 1995) deficient mice, showing that the B cell antigen receptor can be used in different ways. In contrast, the TD, but not the TI-2 immune response is absent in MHC class II (COSGROVE et al. 1991; GRUSBY et al. 1991), CD19 (RICKERT et al. 1995) and CD40/CD40L (RENSHAW et al. 1994; XU et al. 1994; KAWABE et al. 1994; CASTIGLI et al. 1994) deficient mice, all of those mutations affecting T-B cell interaction. In most mutations the defect in production of TD IgG, IgA or IgE isotypes is compensated by higher production of primordial IgM (GRUSBY et al. 1991; KAWABE et al. 1994), i.e., the cells are still activated to the secretion of IgM, but do not switch. Viable TI-1 deficient mutant mice have not been described. In homozygous Oct-2 deficient mice the LPS-induced activation is greatly impaired, but these mice die around birth (CORCORAN et al. 1993; CORCORAN and KARVELAS 1994). The genetic data show that TI-1, TI-2 and TD antigens activate B cells via different signaling pathways and that these pathways are dominated by particular Ig classes.

3.2 B Cell Receptor Complex

In naive surface IgM and IgD expressing B lymphocytes of mouse and humans, Ig class switch recombination can be induced in some, but not all cells by induction of proliferation of the B cells in vitro allowing to define the minimum signal requirements for the recombination (KEARNEY and LAWTON 1975; KEARNEY

Table 1. Genes and DNA sequences involved in murine immunoglobulin class switching

Gene	Reference	Population TH	B2	B1	Proliferation LPS	Anti-μ	Antibody titer in the serum of homozygous mice IgM	IgG3	IgG1	IgG2b	IgG2a	IgE	IgA	Primary immune response TI-1	TI-2	TD	GC	Notes
Complex immunodeficiencies																		
Xid	Khan et al. 1995	↓2	↓13*	↓295*	→	↓16	↓26	↓390	↓27	↓12	37		↓16	-***	-***	=****		)*IgM+IgD+CD5+; ***total Ig's
Btk (Kin)	Khan et al. 1995	=	↓6*	↓98*	→	↓45	↓22	↓29	↓3	=	↓2		↓2	-***	-***	=****		)**** IgM↓; IgG3↓; IgG1=; IgG2b=; IgG2a=;
B cell receptor complex																		
IgD	Nitschke et al. 1993		↓2*				=	=	=	↓5		↑2	=	=**	=***	=	*B220+IgM+; **dex, iv immunization, only IgM; ***DNP-Ov: IgM↑3, IgG2↓3, IgG1↓2; 3 fold increased surface IgM
	Roes and Rajewsky 1993	=	=	=			=	=	=	=	↓3*	↓5	=	=**	=***		*No IgG2a difference in older mice; **dex: IgM=, IgG3↑2, ***NP-CG: IgG1=, IgG2a↑2; delayed affinity maturation
CD45exib6	Kishihara et al. 1993	↓3*	=**				↓4		↓29*		↓21*			↓=**	↓***	-	*Lymph nodes; **IgM+
CD19	Rickert et al. 1995		=	↓5-10	±	→											*Wide Interval; **NP-Ficoll=; dex↓5; ***NP-CG: IgG1↓10-100
Costimulatory signals																		
MHC class II	Cosgrove et al. 1991	20*	=*												-**		*In spleen + lymph nodes; **KLH+Ov: total Ig
	Grusby et al. 1991	↓15*													↑-**		*In lymph nodes; **NP-Ov: IgM↑, all IgG-
	Köntgen et al. 1993	↓15*		=			↑2	↓10-100	↓10	↓4	=				↑**		*In lymph nodes; **levan: only IgM was measured reduced Ig response to SRBC; normal LPS response
CD4	Rahemtulla et al. 1991	-		=													
CD28	Shahinian et al. 1993	=	=*	±			=	↓10	↓20	17							*B220+; total Ig's↓5; defect in lectin induced T cell response: VSV response: IgM=, IgG↓
B7	Freeman et al. 1993	±	±*	±			+	+	↓18	+	↓8	↓44	±	=**			*B220+
C40	Castigli et al. 1994	=	=		=		=	↑2		↓2	+	+	↓3	=*		TM***	-	*TNP-LPS: IgM=, IgG=; **TNP-Ficoll: IgM= IgG=; ***TNP-KLH: IgM-, IgG-; normal LPS/IL-4 switch; defective CD40L induced proliferation + CD40L/IL-4 switch
	Kawabe et al. 1994	±	±		=		=	↑2	↓15	↓3	↓10		↓3	=*	=**	↑-***		*TNP-LPS: IgM=, IgG3↑, IgG2b↓, IgG2a↑; **TNP-Ficoll: IgM=, IgG3=, IgG1=; ***DNP-Ov: IgM↑, all other Ig's: -; IgMbright IgDdull; CD23↓ on mature B cells
CD40L	Xu et al. 1994	=	=				=	↓2	↓24	↓6	↓12	↓>220	↓3	=*	=*	-**	-	*TNP-Ficoll: IgM=, IgG3=; **KLH: IgM↓2; IgG2b↑25, IgA↑;
	Renshaw et al. 1994	=	=				=	↓8	↓8	↓3	↓>1800		↓3	=↑	=↑	-**		*DNP-Ficoll: IgM=, IgG3=, IgG1↑, IgG2b↑25, IgA↑; **TNP-KLH: IgM↓, all other Ig's: -; normal LPS or CD40L induced swich in vitro

Cytokines & cytokine receptors

IL-2	Schorle et al. 1991	=			↑39	↑		=		Inability to induce in vitro IgM secretion upon anti-μ stimulation; reduced polyclonal in vitro T-cell responses
IL-4	Kühn et al. 1991	=		=	↓6	=	=	↓>7	=	*Nb: IgE↓>1000; NP-CG: IgG3↑3, IgG1↓3, IgG2b↑18, IgG2a↑33
	Kopf et al. 1993	=		=	↓20	=	=	–	↓2	↓↑* *Nb: IgG3↑4, IgG1↓12, IgG2b↑4, IgG2a↑3, IgE↓>15000, IgA↑2
IL-6	Kopf et al. 1994	±		±	±	±	±	±	±	=↓* *DNP-Ov: IgG1↓10, IgG2+IgG3↑100-500; block of Th2 cytokine production / *Protein-antigen: IgG↓; VSV: IgM±; IgG↓5-10; defective in mucosal IgA response
IFNγ	Dalton et al. 1993	=*								*B220+; MHC class II ↓ on macrophages
IFNγR	Huang et al. 1993	=*				→				=↓** *IgM+; **TNP-Ov: IgM=, IgG3↓3, IgG1=, IgG2b=, IgG2a↓4; IgE–

Transcription factors

NF-κB	Sha et al. 1995	=*	↑2	↓2	↓13	↓3	↓2	↓48	↓10	↓** *IgM+; **NP-CG: IgM↓6, IgG3↓6, IgG1↓36, IgG2b 36

DNA elements

Jн+Eμ	Gu et al. 1993	→	–							
3'E	Cogne et al. 1994	=		↓100 ↓3	↓4*	↓120 **	↓2			↓2 No switch recombination at sμ, but at sγ1 upon LPS/IL-4 activation / *IgG2b secretion: ↓>200 in LPS ↓>60 in LPS/IL-4 cultures; **IgE secretion: ↓>15 in LPS/IL-4 cultures; defective Iγ3 + Iγ2b switch transcripts upon LPS activation
5'sγ1	Jung et al. 1993	=*		↓50*	=*					=–** *Allotype specific Ig's in heterozygous mice; **NP-CG: IgM=, IgG1↓>750; in vitro LPS/IL-4 induced IgG1+ cells: ↓200
	Lorenz et al. 1995 M. Lorenz and A. Radbruch, manuscript in preparation	=		↓>2500 =	=					=–* *NP-CG: IgM=, IgG1↓>2500
5'sγ2b	Zhang et al. 1993	=	↓2	=	↓120	=				

Table 1. (Continued)

Gene	Population			Proliferation		Antibody titer in the serum of homozygous mice							Primary immune response			GC	Notes
	T_H	B2	B1	LPS	Anti-μ	IgM	IgG3	IgG1	IgG2b	IgG2a	IgE	IgA	TI-1	TI-2	TD		
Iε-Eμ/VDJ																	Sε transcription; LPS Induced IgE↓10-100 compared to LPS/IL-4 induced IgE of control
Iγ1-hMT						=	=	↓>2500	=	=	=	=			=-*		*NP-CG: IgM:=, IgG1↓>2500; Sγ1 transcription; no stable Iγ1 switch transcripts; in vitro LPS/IL-4 induced IgG1+ cells ↓>250
																	M. Lorenz and A. Radbruch, manuscript in preparation
Iγ1-shMT						=	↓70	=	=	=	=	=			=*		*NP-CG: IgM=, IgG1=; Sγ1 transcription; stable Iγ1 switch transcripts; in vitro LPS induced IgG1+ cells in absence of IL-4: ↑130; in vitro LPS/IL-4 induced IgG1+ cells:=
																	Lorenz et al. 1995 M. Lorenz and A. Radbruch, manuscript in preparation

Effects of homozygous mutations as indicated on T_H cell (CD4+), conventional B-2 and B-1 lymphocyte population, their ability to proliferate upon stimulation with lipopolysaccharide (LPS) or cross-linking of IgM receptors (anti-μ), the effect on serum antibody titer of various isotypes in homozygous mice, the ability to mount primary immune responses to TI-1, TI-2 and TD antigens and the effect on germinal center formation (GC) either normal (=) or impaired. The phenotype of mutant mice in immunoglobulin production (Ig) compared to control mice is given as either equal within a factor of 1.5 (=), lower (↓n) or higher (↑>n) by a factor of n, not detectable (↓n) or not detectable and determinable. Parameters mentioned to be unchanged, we indicated by (±). An asterisk (*) refers to a specific note. In some cases, we have tried to calculate the change of Ig levels from figures on our own, if original values were missing.

Nb, Nippostrongylus brasiliensis; VSV, vesicular stomatitis virus; SRBC, sheep red blood cells; NP-, 4-hydroxy-3-nitrophenylacetyl-; DNP-, 2,4-dinitrophenyl-; TNP-, 2,4,6-trinitrophenyl-; Ov, ovalbumin; CG, chicken γ-globulin; KLH, keyhole limpet hemocyanin; dex, α(1–3) dextran; LPS, lipopolysaccharide, iv.

et al. 1976; PERNIS et al. 1977; COUTINHO and FORNI 1982; FINKELMAN and VITETTA 1984). So far, in all those systems for stimulation of naive B cells in vitro, class switching is induced whenever proliferation is induced. Regardless of the mood of stimulation, not all of the cells perform class switching, although in the progeny of about 80% of LPS-activated cells, switched cells do occur (COUTINHO and FORNI 1982). Activation as such, as reflected by transition of cells from G0 to G1 phase of the cell cycle and increased expression of MHC class II molecules is apparently not sufficient. Anti-IgM or anti-IgD both can activate B cells, but do not induce class switching by themselves. In conjunction with other molecules they can, as long as this combination induces proliferation of the B cells. In comparison, surface IgM and IgD both can activate B cells for class switching. IgD-deficient mice, in which the lack of surface IgD is compensated by a threefold increased expression of surface IgM (Nitschke et al. 1993), show no significant change in serum titer of any Ig class and in the immune responses against TI-2 and TD antigens. On the other hand, cross-linking of surface IgD on resting murine B cells by anti-IgD monoclonal antibodies conjugated to dextran in vitro (SNAPPER et al. 1991) or in vivo (THY-PHRONITIS et al. 1993) also induces normal proliferation and, in the presence of cytokines, Ig class switching. Thus, with respect to class switching, IgM and IgD can replace each other.

The signal transducing units of the BCR are Igα and Igβ. Igα mutant mice have a strong, but leaky block in B cell development (TORRES et al. 1995). Immunoglobulin class switching has not been analyzed in these "leaky" B cells. The activation signal transmitted by the BCR seems to be enhanced by the protein tyrosine phosphatase CD45. CD45exon6 deficient mice show a significant reduction of anti-IgM inducible proliferation while retaining their LPS responsiveness (KISHIHARA et al. 1993). A component of the B cell coreceptor complex, CD19, in synergy with CD21, CD81 and Leu13 is involved in B cell activation and in TD responses and in TI-2 responses of B2 cells, but not in TI-2 immune responses of B1 cells, as shown by targeted inactivation of CD19 (RICKERT et al. 1995; Table 1).

3.3 Costimulatory Signals

In CD45exon6 deficient mice the T$_H$ cell population is only slightly altered (KISHIHARA et al. 1993), while in MHC class II (COSGROVE et al. 1991; GRUSBY et al. 1991; KÖNTGEN et al. 1993) and CD4 (RAHEMTULLA et al. 1991) deficient mice few, if any, T$_H$ lymphocytes are found. The frequency of mature conventional (B-2) B cells is not affected (COSGROVE et al. 1991) and those still respond normally to LPS, i.e., TI-1 antigen (RAHEMTULLA et al. 1991; KÖNTGEN et al. 1993). Obviously, TD immune responses are affected. The absence of help from T cells results in a strong and specific serum IgG1 and a slight IgG2b deficiency (COSGROVE et al. 1991). IgG1 and IgG2b are also reduced in the serum of CD28-deficient mice (SHAHINIAN et al. 1993), but not in B7-deficient mice (FREEMAN et al. 1993). The titers of IgM and IgG3, the major antibody classes of TI responses, are not significantly affected in MHC class II-deficient mice which lack T$_H$ cells (Cosgrove et al. 1991). The antibody responses against a number of TI-2 antigens are the same as in control mice (Cosgrove et al. 1991; Köntgen et al. 1993), while the antibody responses against several TD antigens (Cosgrove et al. 1991; Grusby et al. 1991) are completely absent. The defect in generation of TD immune responses is correlated to a defect in germinal center formation (Cosgrove et al. 1991; Xu et al. 1994; Kawabe et al. 1994; Castigli et al. 1994; Rickert et al. 1995). Although it is clear from these data that T$_H$ cells control the Ig class switch to IgG1, the effect is less dramatic than could have been expected. The reduction in titer of IgG1 in the serum could be much higher, when compared with Iγ1-promoter targeted mice (for the hMT-variant, see Lorenz et al. 1995b). Switching to murine IgG1 is apparently not dependent on T cell help, but T cell help can efficiently promote switching to IgG1.

There are many ways in which activated T$_H$ cells can provide direct and indirect help for Ig class switching. Cytokines, mostly secreted, or membrane-bound ligands for BCRs such as the CD40L (TRAP) binding to CD40, CD28 and CTLA-4 binding to B7, and T cell receptor (TCR) and CD4 binding to the B cell MHC class II/antigen peptide-complex, to name just a few, enable T cells to control B cell differentiation.

Since B cell proliferation apparently is a prerequisite for Ig class switching (see below), T cell control of B cell proliferation indirectly also controls B cell class switching. The most prominent interaction in this respect is CD40 and its ligand. This became evident when the genetic defect in the human X-linked hyper IgM syndrome was identified as mutations of the human CD40L gene (Fuleihan et al. 1993; Aruffo et al. 1993; DiSanto et al. 1993; Allen et al. 1993; Korthauer et al. 1993). Patients have little IgG, IgA and IgE but more IgM than healthy people (for review, see Kroczek et al. 1994), reflecting a defect in TD immune responses. CD40- and CD40L- deficient mice (Castigli et al. 1994; Kawabe et al. 1994; Xu et al. 1994; Renshaw et al. 1994) have T$_H$ and conventional B cells at normal frequencies. The lack of CD40 does not abolish LPS-induced proliferation (Castigli et al. 1994; Kawabe et al. 1994) or sensitivity to BCR stimulation with anti-IgM (Kawabe et al. 1994). A significant reduction in IgG1, IgG2b, IgG2a and IgE titers is observed in the serum of mutant mice (Castigli et al. 1994; Kawabe et al. 1994; Renshaw et al. 1994; Xu et al. 1994). Nevertheless, the B cells are endogenously still capable of switching to IgG1 as shown by stimulation of CD40-deficient B cells with LPS and interleukin 4 (IL-4) in vitro (Castigli et al. 1994). The CD40-deficient B cells also switch in vivo to IgG3, the dominant class of TI immune responses. On the other hand, antigen-independent stimulation by CD40 alone of B cells to proliferation seems to be sufficient to class switching at low frequencies (Banchereau et al. 1994). The CD40/CD40L-induced proliferation is enhanced by other stimulatory signals such as ligation of Ox40 (Stuber et al. 1995) and several cytokines. The effect of cytokines like IL-2, IL-4, IL-10, IL-13 and interferon γ (IFNγ) with other signals in stimulation of proliferation is additive (for review, see Banchereau et al. 1994).

3.4 Is Class Switch Recombination Linked to Proliferation?

It is obvious that class switching is not induced by activation of B cells, i.e., transition from G0 to G1 phase. Induction of proliferation seems to be required (Max et al. 1995), although it is not clear whether DNA synthesis as such would not be sufficient. The inhibition of DNA synthesis in LPS-activated murine spleen cells with thymidine or hydroxyurea reduces the frequency of class switching drastically (Severinson-Gronowics et al. 1979). Based on their analysis of presumably related switch recombinations in the murine lymphoma line I.29, Dunnick and Stavnezer have proposed that class switch recombination may be mediated by error-prone DNA repair synthesis of one DNA strand and ligation of the other strand from pre-existing DNA (Dunnick et al. 1989; Dunnick and Stavnezer 1990). Recently, the physiological relevance of these data has been questioned by the demonstration that most class switch recombinations in vivo work by loop-out-and-deletion of intervening DNA (von Schwedler et al. 1990; Iwasato et al. 1990; Matsuoka et al. 1990), supporting the possibility that the I.29 switch events and mutations had resulted from unrelated switch events. Sequence analyses of switch recombination breakpoints recovered from normal B cells have shown small insertions at the recombination sites which would favor a DNA double-strand break and repair mechanism (Winter et al. 1987; Roth et al. 1989). Whether or not class switch recombination requires proliferation or just DNA repair synthesis, it has not yet been possible to dissect class switching from proliferation in terms of signal requirements.

4 Targeting Switch Recombination

4.1 The Role of Switch Regions in Immunoglobulin Class Switching

Switch recombination occurs in activated and proliferating B lymphocytes at frequencies of up to 10% per generation in vitro and probably at even higher frequencies in vivo (Winter et al. 1987). Although aberrant recombinations may happen occasionally and even may be involved in the generation of transforming oncogene translocations, like that of c-myc (Tian and Faust 1987), the overall fidelity of switch recombination is remarkable. The sequences involved, the switch regions, are located 5' of all heavy chain constant region genes, except murine Cδ, which as a consequence is not involved in recombinatorial class switching. Switch regions consist of highly repetitive and GC-rich sequences. In general, GC-rich DNA stretches are known to be highly recombinogenic (Charlesworth 1994), maybe through formation of (CG)*G DNA triplets (Meervelt et al. 1995) or DNA/RNA triplet structures (Reaban and Griffin 1990; Reaban et al. 1994; for review, see Lorenz and Radbruch 1996). We still do not

to tandem repeats of the sγ3, sγ1, and sγ2b switch regions (Wuerffel et al. 1992; Kenter et al. 1993). Many of the sγ3 recombination breakpoints occur within the SNIP/NF-κB-binding site and the binding site of another protein called switch nuclear A protein (SNAP) (Wuerffel et al. 1992). Since the sequence of the transcription factor NF-κB contains a rel domain, which could promote dimerization, it has been speculated that those factor complexes might be involved in the alignment of switch regions for switch recombination.

Another molecule that has been speculated to act as a transcription factor and which may be involved in switch recombination as well, mostly because it binds to switch region sequences, is B-cell specific activator protein (BSAP). The inductor of BSAP (Barberis et al. 1990) is not known. BSAP binds to two distinct sites 5' of the sα region (Waters et al. 1989) and to repetitive sequence motifs of the sμ-region (Xu et al. 1992). The transcription factor BSAP is involved in induction of Iε transcription (Liao et al. 1994) and control of B cell proliferation (Max et al. 1995).

The limitations of conventional gene targeting strategies become obvious in the analysis of the role of transcription factors in switch recombination. Since most of the candidate transcription factors, identified by their switch cytokine induced activation or their binding to switch regions or 5' thereof are pleiotropic, their knockout is lethal or at least obstructs early B cell development. An example is BSAP, which is essential for B cell development. No pre-B cells, mature B cells, or plasma cells can be detected in Pax5/BSAP deficient mice, making it impossible to analyze class switching in these mutant mice (Urbanek et al. 1994). PAX5/BSAP also seems to play a key role in mouse development, since its targeted deletion leads to neonatal mortality (Urbanek et al. 1994). Thus the role of BSAP in regulation of Ig class switch recombination remains obscure after all.

Targeting of the transcription factors E2A (Bain et al. 1994; Zhuang et al. 1994) and Oct-2 (Corcoran et al. 1993), which also might be involved in regulation of switch transcription, leads to neonatal and perinatal mortality, respectively. Oct-2-deficient mice show normal numbers of pre-B cells, but no mature B cells. The LPS-induced Ig production is greatly impaired in Oct-2-deficient B cells (Corcoran et al. 1993; Corcoran and Karvelas 1994).

4.5 DNA Sequences Involved in Targeting of Switch Recombination

While the targeted knockout of genes for transcription factors has not contributed much to understanding their role and the role of switch transcription for the control of switch recombination, the mutational analysis of the DNA sequences involved in switch transcription has established the functional dependency of switch recombination on switch transcription. The deletion of entire promoter regions for switch transcription such as the Iγ1 (Jung et al. 1993) or Iγ2b (Zhang et al. 1993) cytokine-dependent promoter regions of the mouse, lead to a severe and selective IgG1 or IgG2b deficiency, respectively. Little, if any, IgG1 is detected in the Iγ1 knockout mice. In the mutant B cells, the switch promoter controls targeting of recombination to the respective adjacent switch region and only on that particular IgH locus, i.e., in *cis*.

Deletion of the JH gene segments and the Ig heavy chain intron enhancer, located just 5' of the sμ switch region and containing the promoter for Iμ transcripts (LENNON and PERRY 1985; SU and KADESCH 1990) results in a drastic reduction of switch recombination involving sμ on the mutant chromosome (GU et al. 1993). If class switching from IgM to IgG1 is induced by LPS and IL-4, the sγ1 switch region performs switch recombination with itself. These intra-switch region recombinations seem to be a characteristic feature of switching cells and intra sμ recombinations had previously been observed in switching normal B cells at high frequency (WINTER et al. 1987).

While the IgH "intron"-enhancer apparently is critical for switch recombination of the sμ switch region, the murine 3' IgH-enhancer, located 12.5 kb downstream of the Cα-membrane exon (DARIAVACH et al. 1991; LIEBERSON et al. 1991), acts as an enhancer for switch recombination to various classes. Its targeted mutation inhibits switching to IgG2a, IgG2b, IgG3 and IgE in vitro and, to a lesser extent of IgG2b in vivo (COGNE et al. 1994). It is not quite clear whether the enhancement in Ig production is due to an enhancement of switch transcription or whether other effects may be involved. Inactivation of the 3' enhancer abolishes the formation of Iγ2b and Iγ3 switch transcripts in B cells stimulated with LPS, but has no effect on the production of Iμ and Iγ1 switch transcripts in LPS plus IL-4 activated B cells, in accordance with the effect on switch recombination (COGNE et al. 1994). Class switching to IgG1 is only marginally affected in the absence of a functional 3' enhancer.

4.6 Switch Transcripts and Switch Recombination

The correlation between switch transcription and switch recombination in terms of time, induction signals, specificity and requirement of sequences 5' of switch regions is well documented, as discussed above. The functional interrelation between transcription and recombination is far less understood. Initially, it had not even been clear, whether switch transcription was just an epiphenomenon, or required to "open" the DNA of switch regions, or whether the switch transcript would participate in switch recombination. Since transcription induces negative supercoils in front of the RNA polymerase complex, temporary RNA-DNA hetero-duplices and single-stranded DNA, switch transcription could well target a particular switch region for recombination. It has been shown that transcription supports DNA repair (for review, see HANAWALT et al. 1994). Unfortunately, participation of excision repair cross complementing protein (ERCC-1) (McWHIR et al. 1993) in Ig class switch recombination remains obscure since its targeted mutation results in a lethal phenotype. However, for switch recombination it is already clear that transcription of switch regions is not sufficient to target that respective switch region for recombination. Replacing the Iε- or Iγ1-promoters responsible for switch transcription with a V$_H$ gene promoter or a human metallothionein promoter, respectively, results in drastic reduction of switch recombination frequencies, despite considerable transcription (BOTTARO et al. 1994; LORENZ et al. 1995b). In these mutant B cells, however, the resulting transcripts cannot be processed into spliced switch transcripts. If replacement of the Iγ1 promoter by a human metallothionein promoter is performed in such a way that artificial switch transcript are generated by splicing, the heterologous promoter can now target switch recombination to sγ1 (LORENZ et al. 1995b). This is in accordance with recent data that the direction of transcription through switch regions is important (DANIELS and LIEBER 1995). The replacement mutations have also shown that the endogenous I-exon sequences, at least the Iγ1 and Iα exon sequences, as such are not involved in targeting of class switch recombination because switch transcripts without or with largely reduced endogenous I-exon sequences can target switch recombination (LORENZ et al. 1995a,b; HARRIMAN and ROGERS 1995). Putative switch peptides encoded by physiological switch transcripts do not play a role in targeting switch recombination. Although these recent data strongly suggest that switch transcripts as such or the process of their generation are involved in switch recombination, it remains obscure how they can do so. But it seems clear by now that cytokine-induced switch transcripts target switch recombination to particular switch regions. Switch recombination is controlled on the level of switch transcription by cytokines.

5 Down-Regulation of Switch Recombination

Immunoglobulin class switch recombination is probably not a single recombinatorial event. Most likely, multiple recombinations occur within and between targeted switch regions, as is obvious from the analysis of switch recombinations in individual activated B cells (WINTER et al. 1987), inactivation of sμ for switch recombination with subsequent intra sγ1 recombinations (GU et al. 1993), and the analysis of sequential switching from IgM to IgE via IgG1 (SIEBENKOTTEN et al. 1992). The actual switch recombination leads to excised switch circles (VON SCHWEDLER et al. 1990; IWASATO et al. 1990; MATSUOKA et al. 1990) and chimeric genomic switch regions. Comparison of recombination breakpoints in switch circles, i.e., the excised DNA, and genomic switch recombination products reveals that sμ switch recombination break points are located within the repetitive sμ sequences of switch circles (VON SCHWEDLER et al. 1990; IWASATO et al. 1990, MATSUOKA et al. 1990), but 5' thereof on the chromosome (SZUREK et al. 1985; WINTER et al. 1987), indicating additional recombinations on the chromosome after the initial recombination, which has generated the switch circle (ZHANG et al. 1995). In switched cells, the new chimeric switch region can be targeted by the Iμ switch promoter for further switch recombinations (LI et al. 1994). Although not obligatory, sequential switch recombination has been shown to be the major pathway for generation of murine IgE (SIEBENKOTTEN et al. 1992).

Immunoglobulin class switch recombination seems to be turned off at the plasma cell stage. In hybrid-hybridomas of polyclonally activated B cells and PC140 hybridoma cells, the frequency of class switching is not enhanced for PC140 IgH loci, but rather inhibited for the IgH loci of the LPS blasts (KLEIN and RADBRUCH 1994). An inhibitor may exist in plasma cells, down-regulating switch recombination. In any case, the frequency of switch recombination in immortalized plasma blasts or plasma cells such as hybridoma or myeloma cells is very low, and in many aspects does not resemble physiological class switching, e.g., often sequences other than switch regions are involved (reviewed in SABLITZKY et al. 1982). Also, switch recombination substrates are only poorly recombined in myeloma cells (OTT et al. 1987; OTT and MARCU 1989).

References

Allen RC, Armitage RJ, Conley ME, Rosenblatt H, Jenkins NA, Copeland NG, Bedell MA, Edelhoff S, Disteche CM, Simoneaux DK, et al (1993) CD40 ligand gene defects responsible for X-linked hyper-IgM syndrome. Science 259:990–993

Aruffo A, Farrington M, Hollenbaugh D, Li X, Milatovich A, Nonoyama S, Bajorath J, Grosmaire LS, Stenkamp R, Neubauer M, et al (1993) The CD40 ligand, gp39, is defective in activated T cells from patients with X-linked hyper-IgM syndrome. Cell 72:291–300

Bain G, Maandag EC, Izon DJ, Amsen D, Kruisbeek AM, Weintraub BC, Krop I, Schlissel MS, Feeney AJ, van Roon M, et al (1994) E2A proteins are required for proper B cell development and initiation of immunoglobulin gene rearrangements. Cell 79:885–892

Banchereau J, Bazan F, Blanchard D, Briere F, Galizzi JP, van-Kooten C, Liu YJ, Rousset F, Saeland S (1994) The CD40 antigen and its ligand. Annu Rev Immunol 12:881–922

Barberis A, Widenhorn K, Vitelli L, Busslinger M (1990) A novel B-cell lineage-specific transcription factor present at early but not late stages of differentiation. Genes Dev 4:849–859

Bottaro A, Lansford R, Xu L, Zhang J, Rothman P, Alt FW (1994) S region transcription per se promotes basal IgE class switch recombination but additional factors regulate the efficiency of the process. EMBO J 13:665–674

Brys A, Maizels N (1994) LR1 regulates c-myc transcription in B-cell lymphomas. Proc Natl Acad Sci USA 91:4915–4919

Burrows P, LeJeune M, Kearney JF (1979) Evidence that murine pre-B cells synthesise mu heavy chains but no light chains. Nature 280:838–840

Burrows PD, Beck GB, Wabl MR (1981) Expression of mu and gamma immunoglobulin heavy chains in different cells of a cloned mouse lymphoid line. Proc Natl Acad Sci USA 78:564–568

Burrows PD, Beck-Engeser GB, Wabl MR (1983) Immunoglobulin heavy-chain class switching in a pre-B cell line is accompanied by DNA rearrangement. Nature 306:243–246

Castigli E, Alt FW, Davidson L, Bottaro A, Mizoguchi E, Bhan AK, Geha RS (1994) CD40-deficient mice generated by recombination-activating gene-2- deficient blastocyst complementation. Proc Natl Acad Sci USA 91:12135–12139

Charlesworth B (1994) Genetic recombination. Patterns in the genome. Curr Biol 4584:182–184

Coffman RL, Lebman DA, Rothman P (1993) Mechanism and regulation of immunoglobulin isotype switching. Adv Immunol 54:229–270

Cogne M, Lansford R, Bottaro A, Zhang J, Gorman J, Young F, Cheng HL, Alt FW (1994) A class switch control region at the 3' end of the immunoglobulin heavy chain locus. Cell 77:737–747

Corcoran LM, Karvelas M (1994) Oct-2 is required early in T cell-independent B cell activation for G1 progression and for proliferation. Immunity 1:635–645

Corcoran LM, Karvelas M, Nossal GJ, Ye ZS, Jacks T, Baltimore D (1993) Oct-2, although not required for early B-cell development, is critical for later B-cell maturation and for postnatal survival. Genes Dev 7:570–582

Cosgrove D, Gray D, Dierich A, Kaufman J, Lemeur M, Benoist C, Mathis D (1991) Mice lacking MHC class II molecules. Cell 66:1051–1066

Coutinho A, Forni L (1982) Intraclonal diversification in immunoglobulin isotype secretion: an analysis of switch probabilities. EMBO J 1:1251–1257

Dalton DK, Pitts-Meek S, Keshav S, Figari IS, Bradley A, Stewart TA (1993) Multiple defects of immune cell function in mice with disrupted interferon-gamma genes [see comments]. Science 259:1739–1742

Daniels GA, Lieber MR (1995) Strand specificity in the transcriptional targeting of recombination at immunoglobulin switch sequences. Proc Natl Acad Sci USA 92:5625–5629

Dariavach P, Williams GT, Campbell K, Pettersson S, Neuberger MS (1991) The mouse IgH 3' -enhancer. Eur J Immunol 21:1499–1504

DiSanto JP, Bonnefoy JY, Gauchat JF, Fischer A, de Saint Basile G (1993) CD40 ligand mutations in x-linked immunodeficiency with hyper-IgM. Nature 361:541–543

Dunnick W, Stavnezer J (1990) Copy choice mechanism of immunoglobulin heavy-chain switch recombination. Mol Cell Biol 10:397–400

Dunnick W, Wilson M, Stavnezer J (1989) Mutations, duplication, and deletion of recombined switch regions suggest a role for DNA replication in the immunoglobulin heavy-chain switch. Mol Cell Biol 9:1850–1856

Finkelman FD, Vitetta ES (1984) Role of surface immunoglobulin in B lymphocyte activation. Fed Proc 43:2624–2632

Francis DA, Karras JG, Ke XY, Sen R, Rothstein TL (1995) Induction of the transcription factors NF-kappa B, AP-1 and NF-AT during B cell stimulation through the CD40 receptor. Int Immunol 7:151–161

Freeman GJ, Borriello F, Hodes RJ, Reiser H, Hathcock KS, Laszlo G, McKnight AJ, Kim J, Du L, Lombard DB, et al (1993) Uncovering of functional alternative CTLA-4 counter-receptor in B7-deficient mice. Science 262:907–909

Fuleihan R, Ramesh N, Loh R, Jabara H, Rosen RS, Chatila T, Fu SM, Stamenkovic I, Geha RS (1993) Defective expression of the CD40 ligand in X chromosome-linked immunoglobulin deficiency with normal or elevated IgM. Proc Natl Acad Sci USA 90:2170–2173

Grusby MJ, Johnson RS, Papaioannou VE, Glimcher LH (1991) Depletion of CD4+ T cells in major histocompatibility complex class II-deficient mice. Science 253:1417–1420

Gu H, Zou YR, Rajewsky K (1993) Independent control of immunoglobulin switch recombination at individual switch regions evidenced through Cre-loxP-mediated gene targeting. Cell 73:1155–1164

Hanawalt PC, Donahue BA, Sweder KS (1994) Repair and transcription. Collision or collusion? Curr Biol 4:518–521

Harriman G, Rogers P (1995) TGF-beta-independent IgA class switch in mice with targeted deletion of the Ialpha exon. The 9th congress of immunology, San Francisco, 23.7.1995–29.7.1995

Harriman W, Volk H, Defranoux N, Wabl M (1993) Immunoglobulin class switch recombination. Annu Rev Immunol 11:361–384

Hou J, Schindler U, Henzel WJ, Ho TC, Brasseur M, McKnight SL (1994) An interleukin-4-induced transcription factor: IL-4 Stat. Science 265:1701–1706

Huang S, Hendriks W, Althage A, Hemmi S, Bluethmann H, Kamijo R, Vilcek J, Zinkernagel RM, Aguet M (1993) Immune response in mice that lack the interferon-gamma receptor . Science 259:1742–1745

Isakson PC, Pure E, Vitetta ES, Krammer PH (1982) T cell-derived B cell differentiation factor(s). Effect on the isotype switch of murine B cells. J Exp Med 155:734–748

Iwasato T, Shimizu A, Honjo T, Yamagishi H (1990) Circular DNA is excised by immunoglobulin class switch recombination. Cell 62:143–149

Jung S, Rajewsky K, Radbruch A (1993) Shutdown of class switch recombination by deletion of a switch region control element. Science 259:984–987

Kawabe T, Naka T, Yoshida K, Tanaka T, Fujiwara H, Suematsu S, Yoshida N, Kishimoto T, Kikutani H (1994) The immune responses in CD40-deficient mice: impaired immunoglobulin class switching and germinal center formation. Immunity 1:167–178

Kearney JF, Lawton AR (1975) B lymphocyte differentiation induced by lipopolysaccharide. I. Generation of cells synthesizing four major immunoglobulin classes. J Immunol 115:671–676

Kearney JF, Cooper MD, Lawton AR (1976) B cell differentiation induced by lipopolysaccharide. IV. Development of immunoglobulin class restriction in precursors of IgG-synthesizing cells. J Immunol 117:1567–1572

Kenter AL, Wuerffel R, Sen R, Jamieson CE, Merkulov GV (1993) Switch recombination breakpoints occur at nonrandom positions in the S gamma tandem repeat. J Immunol 151:4718–4731

Kerner J, Appleby M, Mohr R, Chien S, Rawlings D, Mailszewski C, Witte O, Perlmutter R (1995) Impaired expansion of mouse B cell progenitors lacking Btk. Immunity 3:301–312

Khan W, Alt F, Gerstein R, Malynn B, Larsson I, Rathbun G, Davidson L, Mller S, Kantor A, Herzenberg L, Rosen F, Sideras P (1995) Defective B cell development and function in Btk-deficient mice. Immunity 3:283–299

Kishihara K, Penninger J, Wallace VA, Kundig TM, Kawai K, Wakeham A, Timms E, Pfeffer K, Ohashi PS, Thomas ML, et al (1993) Normal B lymphocyte development but impaired T cell maturation in CD45- exon6 protein tyrosine phosphatase-deficient mice. Cell 74:143–156

Klein S, Radbruch A (1994) Inhibition of class switch recombination in plasma cells. Cell Immunol 157:106–117

Kntgen F, Suss G, Stewart C, Steinmetz M, Bluethmann H (1993) Targeted disruption of the MHC class II Aa gene in C57BL/6 mice. Int Immunol 5:957–964

Kopf M, Le Gros G, Bachmann M, Lamers MC, Bluethmann H, Kohler G (1993) Disruption of the murine IL-4 gene blocks Th2 cytokine responses. Nature 362:245–248

Kopf M, Baumann H, Freer G, Freudenberg M, Lamers M, Kishimoto T, Zinkernagel R, Bluethmann H, Kohler G (1994) Impaired immune and acute-phase responses in interleukin-6-deficient mice. Nature 3681:339–342

Korthauer U, Graf D, Mages HW, Briere F, Padayachee M, Malcolm S, Ugazio AG, Notarangelo LD, Levinsky RJ, Kroczek RA (1993) Defective expression of T-cell CD40 ligand causes X-linked immunodeficiency with hyper-IgM. Nature 361:539–541

Kroczek RA, Graf D, Brugnoni D, Giliani S, Korthuer U, Ugazio A, Senger G, Mages HW, Villa A, Notarangelo LD (1994) Defective expression of CD40 ligang on T cells causes "X-linked immunodeficiency with hyper-IgM (HIGM1)". Immunol Rev 138:39–59

Kubagawa H, Gathings WE, Levitt D, Kearney JF, Cooper MD (1982) Immunoglobulin isotype expression of normal pre-B cells as determined by immunofluorescence. J Clin Immunol 2:264–269

Khn R, Rajewsky K, Muller W (1991) Generation and analysis of interleukin-4 deficient mice. Science 254:707–710

Kulkarni AB, Huh CG, Becker D, Geiser A, Lyght M, Flanders KC, Roberts AB, Sporn MB, Ward JM, Karlsson S (1993) Transforming growth factor beta 1 null mutation in mice causes excessive inflammatory response and early death. Proc Natl Acad Sci USA 90:770–774

Layton JE, Vitetta ES, Uhr JW, Krammer PH (1984) Clonal analysis of B cells induced to secrete IgG by T cell-derived lymphokine(s). J Exp Med 160:1850–1863

Lennon GG, Perry RP (1985) C mu-containing transcripts initiate heterogeneously within the IgH enhancer region and contain a novel 5' -nontranslatable exon. Nature 318:475–478

Leung H, Maizels N (1992) Transcriptional regulatory elements stimulate recombination in extrachromosomal substrates carrying immunoglobulin switch-region sequences. Proc Natl Acad Sci USA 89:4154–4158

Li SC, Rothman PB, Zhang J, Chan C, Hirsh D, Alt FW (1994) Expression of I mu-C gamma hybrid germline transcripts subsequent to immunoglobulin heavy chain class switching. Int Immunol 6:491–497

Liao F, Birshtein BK, Busslinger M, Rothman P (1994) The transcription factor BSAP (NF-HB) is essential for immunoglobulin germ-line epsilon transcription. J Immunol 152:2904–2911

Lieberson R, Giannini SL, Birshtein BK, Eckhardt LA (1991) An enhancer at the 3' end of the mouse immunoglobulin heavy chain locus. Nucleic Acids Res 19:933–937

Lorenz M, Radbruch A (1996) Immunoglobulin class switching. In: Snapper CM (ed) Cytokine regulation of humoral immunity: basic and clinical aspects. Wiley, Chichester

Lorenz M, Jung S, Radbruch A (1995a) How cytokines control immunoglobulin class switching. Behring Institute Mitteilungen 96:97–102

Lorenz M, Jung S, Radbruch A (1995b) Switch transcripts in immunoglobulin class switching. Science 267:1825–1828

Matsuoka M, Yoshida K, Maeda T, Usuda S, Sakano H (1990) Switch circular DNA formed in cytokine-treated mouse splenocytes: evidence for intramolecular DNA deletion in immunoglobulin class switching. Cell 62:135–142

Max EE, Wakatsuki Y, Neurath MF, Strober W (1995) The role of BSAP in immunoglobulin isotype switching and B-cell proliferation. Curr Top Microbiol Immunol 194:449–458

McWhir J, Selfridge J, Harrison DJ, Squires S, Melton DW (1993) Mice with DNA repair gene (ERCC-1) deficiency have elevated levels of p53, liver nuclear abnormalities and die before weaning. Nat Genet 5:217–224

Meervelt L, Vlieghe D, Dautant A, Gallois B, Precigoux G, Kennard O (1995) High-resolution structure of a helix forming (C.G)*G base triplet. Nature 374:742–744

Nitschke L, Kosco MH, Kohler G, Lamers MC (1993) Immunoglobulin D-deficient mice can mount normal immune responses to thymus-independent and -dependent antigens. Proc Natl Acad Sci USA 90:1887–1891

Ott DE, Marcu KB (1989) Molecular requirements for immunoglobulin heavy chain constant region gene switch-recombination revealed with switch-substrate retroviruses. Int Immunol 1:582–591

Ott DE, Alt FW, Marcu KB (1987) Immunoglobulin heavy chain switch region recombination within a retroviral vector in murine pre-B cells. EMBO J 6:577–584

Pernis B, Forni L, Luzzati AL (1977) Synthesis of multiple immunoglobulin classes by single lymphocytes. Cold Spring Harb Symp Quant Biol 1:175–183

Radbruch A, Muller W, Rajewsky K (1986) Class switch recombination is IgG1 specific on active and inactive IgH loci of IgG1-secreting B-cell blasts. Proc Natl Acad Sci USA 83:3954–3957

Rahemtulla A, Fung-Leung WP, Schilham MW, Kundig TM, Sambhara SR, Narendran A, Arabian A, Wakeham A, Paige CJ, Zinkernagel RM, et al (1991) Normal development and function of CD8+ cells but markedly decreased helper cell activity in mice lacking CD4. Nature 353:180–184

Reaban ME, Griffin JA (1990) Induction of RNA-stabilized DNA conformers by transcription of an immunoglobulin switch region. Nature 348:342–344

Reaban ME, Lebowitz J, Griffin JA (1994) Transcription induces the formation of a stable RNA.DNA hybrid in the immunoglobulin alpha switch region. J Biol Chem 269:21850–21857

Renshaw BR, Fanslow W3, Armitage RJ, Campbell KA, Liggitt D, Wright B, Davison BL, Maliszewski CR (1994) Humoral immune responses in CD40 ligand-deficient mice. J Exp Med 180:1889–1900

Rickert RC, Rajewsky K, Roes J (1995) Impairment of T-cell-dependent B-cell responses and B-1 cell development in CD19-deficient mice. Nature 376:352–355

Roes J, Rajewsky K (1993) Immunoglobulin D (IgD)-deficient mice reveal an auxiliary receptor function for IgD in antigen-mediated recruitment of B cells. J Exp Med 177:45–55

Roth DB, Chang XB, Wilson JH (1989) Comparison of filler DNA at immune, nonimmune, and oncogenic rearrangements suggests multiple mechanisms of formation. Mol Cell Biol 9:3049–3057

Sablitzky F, Radbruch A, Rajewsky K (1982) Spontaneous immunoglobulin class switching in myeloma and hybridoma cell lines differs from physiological class switching. Immunol Rev 67:59–72

Schindler C, Kashleva H, Pernis A, Pine R, Rothman P (1994) STF-IL-4: a novel IL-4-induced signal transducing factor. EMBO J 13:1350–1356

Schorle H, Holtschke T, Hunig T, Schimpl A, Horak I (1991) Development and function of T cells in mice rendered interleukin-2 deficient by gene targeting. Nature 352:621–624

Severinson Gronowics E, Doss C, Scrder J (1979) Activation to IgG secretion by lipopolysaccharide requires several proliferation cycles. J. of Immunol. 123:2057–2062

Sha WC, Liou HC, Tuomanen EI, Baltimore D (1995) Targeted disruption of the p50 subunit of NF-kappa B leads to multifocal defects in mmune responses. Cell 80:321–330

Shahinian A, Pfeffer K, Lee KP, Kundig TM, Kishihara K, Wakeham A, Kawai K, Ohashi PS, Thompson CB, Mak TW (1993) Differential T cell costimulatory requirements in CD28-deficient mice. Science 261:609–612

Shull MM, Ormsby I, Kier AB, Pawlowski S, Diebold RJ, Yin M, Allen R, Sidman C, Proetzel G, Calvin D, et al (1992) Targeted disruption of the mouse transforming growth factor-beta 1 gene results in multifocal inflammatory disease. Nature 359:693–699

Siebenkotten G, Radbruch A (1995) Towards a molecular understanding of immunoglobulin class switching. Immunologist 3:141–145

Siebenkotten G, Esser C, Wabl M, Radbruch A (1992) The murine IgG1/IgE class switch program. Eur J Immunol 22:1827–1834

Snapper CM, Pecanha LM, Levine AD, Mond JJ (1991) IgE class switching is critically dependent upon the nature of the B cell activator, in addition to the presence of IL-4. J Immunol 147:1163–1170

Stuber E, Neurath M, Calderhead D, Fell HP, Strober W (1995) Cross-linking of OX40 ligand, a member of the TNF/NGF cytokine family, induces proliferation and differentiation in murine splenic B cells. Immunity 2:507–521

Su LK, Kadesch T (1990) The immunoglobulin heavy-chain enhancer functions as the promoter for I mu sterile transcription. Mol Cell Biol 10:2619–2624

Szurek P, Petrini J, Dunnick W (1985) Complete nucleotide sequence of the murine gamma 3 switch region and analysis of switch recombination sites in two gamma 3-expressing hybridomas. J Immunol 135:620–626

Thyphronitis G, Katona IM, Gause WC, Finkelman FD (1993) Germline and productive C epsilon gene expression during in vivo IgE responses. J Immunol 151:4128–4136

Tian SS, Faust C (1987) Rearrangement of rat immunoglobulin E heavy-chain and c-myc genes in the B-cell immunocytoma IR162. Mol Cell Biol 7:2614–2619

Torres R, Flaswinkel H, Reth M, Rajewsky K (1995) Analysis of the Ig-alpha immunoreceptor tyrosine-based activation motif by gene targeting. The 9th international congress of immunology, San Francisco, 23.7.1995–29.7.1995

Urbanek P, Wang ZQ, Fetka I, Wagner EF, Busslinger M (1994) Complete block of early B cell differentiation and altered patterning of the posterior midbrain in mice lacking Pax5/BSAP. Cell 79:901–912

Vercelli D, Geha RS (1992) Regulation of isotype switching. Curr Opin Immunol 4:794–797

von Schwedler U, Jack HM, Wabl M (1990) Circular DNA is a product of the immunoglobulin class switch rearrangement. Nature 345:452–456

Waters SH, Saikh KU, Stavnezer J (1989) A B-cell-specific nuclear protein that binds to DNA sites 5' to immunoglobulin S alpha tandem repeats is regulated during differentiation. Mol Cell Biol 9:5594–5601

Williams M, Maizels N (1991) LR1, a lipopolysaccharide-responsive factor with binding sites in the immunoglobulin switch regions and heavy-chain enhancer. Genes Dev 5:2353–2361

Winter E, Krawinkel U, Radbruch A (1987) Directed Ig class switch recombination in activated murine B cells. EMBO J 6:1663–1671

Wuerffel R, Jamieson CE, Morgan L, Merkulov GV, Sen R, Kenter AL (1992) Switch recombination breakpoints are strictly correlated with DNA recognition motifs for immunoglobulin S gamma 3 DNA-binding proteins. J Exp Med 176:339–349

Xu J, Foy TM, Laman JD, Elliott EA, Dunn JJ, Waldschmidt TJ, Elsemore J, Noelle RJ, Flavell RA (1994) Mice deficient for the CD40 ligand. Immunity 1:423–431

Xu L, Kim MG, Marcu KB (1992) Properties of B cell stage specific and ubiquitous nuclear factors binding to immunoglobulin heavy chain gene switch regions. Int Immunol 4:875–887

Zhang J, Bottaro A, Li S, Stewart V, Alt FW (1993) A selective defect in IgG2b switching as a result of targeted mutation of the I gamma 2b promoter and exon. EMBO J 12:3529–3537

Zhang K, Cheah HK, Saxon A (1995) Secondary deletional recombination of rearranged switch region in Ig isotype-switched B cells. A mechanism for isotype stabilization. J Immunol 154:2237–2247

Zhuang Y, Soriano P, Weintraub H (1994) The helix-loop-helix gene E2A is required for B cell formation. Cell 79:875–884

Ziegler-Heitbrock HW, Pechumer H, Petersmann I, Durieux JJ, Vita N, Labeta MO, Strobel M (1994) CD14 is expressed and functional in human B cells. Eur J Immunol 2436:1937–1940

Transcription Targets Recombination at Immunoglobulin Switch Sequences in a Strand-Specific Manner

Gregory A. Daniels and Michael R. Lieber

1 Introduction

Immunoglobulin (Ig) heavy chain gene assembly can be divided into two developmentally and mechanistically distinct phases during B cell maturation. The initial, antigen-independent phase occurs primarily in the bone marrow and is catalyzed by the V(D)J recombination activity. The second, primarily antigen-dependent phase involves a second type of gene rearrangement termed class switch recombination. A major mechanism of isotype switching is directed DNA recombination (Davis et al. 1980; Schwedler et al. 1990). Class switch recombination takes place in the peripheral lymphoid tissues and replaces the initial constant regions, $C\mu$ and $C\delta$, with any of several downstream constant regions (for reviews, see Coffman et al. 1993; Gritzmacher 1989). By changing antibody class, or isotype, the same antigen specificity can be associated with a variety of effector functions.

Division of Molecular Oncology, Departments of Pathology, Medicine, and Biochemistry, 660 S. Euclid Ave., Box 8118, Washington University School of Medicine, St. Louis, MO 63110, USA

Two major models for the regulation of class switch recombination have been proposed. One model hypothesizes that different recombinases exist for each switch region, explaining why DNA sequences of the switch regions are different from one another and the commitment of some cells to a limited repertoire of switch events (Davis et al. 1980; Jäck et al. 1988). The second proposal assumes a more general switch recombination mechanism that functions in all switch events but is differentially targeted in some fashion (Stavnezer et al. 1984). Differential targeting of the switch regions might be achieved in any number of ways. Maintaining the switch region DNA in an inaccessible state requiring specific activation is one possibility (Yancopoulos et al. 1986). Prior to recombination to a given switch region, induction of promoter activity, protein binding, and DNase I hypersensitivity occur in the regions 5' of the switch regions (upstream control regions). The temporal correlation between transcriptional and recombinational activation raises the possibility that transcription of switch regions either modulates DNA accessibility or is an indicator of locus accessibility (Lutzker and Alt 1989; Stavnezer-Nordgren and Sirlin 1986). Though transcriptional elongation has been demonstrated to affect the position of individual nucleosomes (Clark and Felsenfeld 1992), effects of transcription on higher order chromatin structure are undefined. One can imagine transcription generating locus targeting for switch recombination in other ways than chromatin effects. For example, the transcripts themselves may directly target the locus (Reaban and Griffin 1990).

From a regulatory standpoint, several studies have attempted to address the importance of the transcriptional control region and the I exon upstream of the switch regions (Jung et al. 1993; Xu et al. 1993; Zhang et al. 1993; Lepse et al. 1994). In one of these (Xu et al. 1993), the Eμ/Vh promoter was inserted in place of the endogenous control region for Sε. These cells were found to undergo a noninducible but low level of switch recombination. Jung et. al. (1993) did the reciprocal experiment by completely removing the upstream control region but leaving Sγ1 intact. The absence of this region greatly diminished recombination. Similar results were obtained for Sγ2b (Zhang et al. 1993). These studies support a positive regulatory role for the upstream control regions but do not dissect out the mechanistic role of transcription from the I exon or other functions of these regulatory regions. In all three cases, enhancers or other implied control elements were also manipulated. It remains a possibility that transcription has either a causal role in recombination or is a temporally related, but causally unrelated, consequence of locus opening.

Previous work has used artificial substrates to study class switch recombination (Leung and Maizels 1992, 1994; Ott et al. 1987). Using both endogenous control regions and heterologous promoter/enhancers (Leung and Maizels 1992), the data were consistent with a causal link between transcriptional regulatory elements and recombination, though without directly linking recombination to transcriptional elongation. In fact, subsequent use of these recombination substrates reported that an enhancer component of the upstream activating region is important for recombination and not transcription itself (LEUNG and MAIZELS 1994).

In order to ask specific mechanistic questions, we have recapitulated switch recombination on extrachromosomal substrates (DANIELS and LIEBER 1995). Our studies have found a causal link between transcription and recombination and suggest that the transcript targets recombination to switch region sequences.

2 Substrate and Assay Design

We have designed a switch sequence cellular assay (DANIELS and LIEBER 1995) similar to the autonomously replicating minichromosome system developed for the study of V(D)J recombination (HESSE et al. 1987; LIEBER et al. 1987). The switch sequence-bearing episomes can be introduced into cells via a variety of transfection methods and recovered up to 60 h later for analysis (Fig. 1). In these substrates, the tRNA gene, *supF*, complements an amber mutation in the *Escherichia coli* strain MLB7070 giving rise to blue colonies on X-GAL plates. Sequences of interest are placed in regions A and B, which flank *supF*. Recombination is scored by the deletion of *supF* and is reported

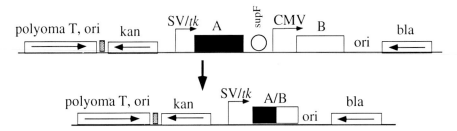

Fig. 1. Switch sequence minichromosome substrates.In the substrate (*top line*), the tRNA gene, *supF*, complements an amber mutation in the *E. coli* strain MLB7070, giving rise to blue colonies on X-GAL plates. Switch or control sequences are placed in regions A and B, which flank *supF*. The substrates are transfected into murine hematopoietic cell lines, incubated for specified times, and the plasmid DNA is harvested and transformed into *E. coli* for genetic assay of product (*bottom line*). Recombination is scored by the deletion of *supF* and is reported as the change in ratio of replicated recombinant product molecules (which give rise to white colonies) to total replicated molecules (blue and white colonies) from 24 to 48 h post transfection. All values are reported as the average slope (with a standard error) of at least three independent transfections (seeSect. 2). The kan and *bla* segments are the kanamycin (*Kn*) and ampicillin (*Ap*) resistance genes, respectively. The polyoma large T antigen and polyoma origin (*polyoma T, ori*) allow replication of the substrate in murine cells. Between the polyoma large T and the kanamycin gene is the gastrin transcriptional terminator (*filled rectangle*). SV/*tk* and hCMV are constitutive eukaryotic promoters oriented in the direction of regions A and B. Ori is the prokaryotic origin of pBR322. *Arrows* represent the direction of transcription

as the change in ratio of replicated recombinant molecules (white colonies) to total replicated molecules (blue and white colonies) from 24 to 48 h post transfection. All values are reported as the average slope (with a standard error) of at least three independent transfections. Because transcriptional activation has been associated with class switch recombination, we have bypassed tissue-specific regulation of these promoters by the use of high level constitutive promoters. Both the SV/*tk* and hCMV promoters function in a wide variety of cells and are active in all of the cell lines used in this study (data not shown). In order to facilitate the analysis of recombinants, a second prokaryotic selection marker was inserted upstream of the promoter for region A. Therefore, by selecting for recombinants which preserve both the β-lactamase and kanamycin resistance genes, we can readily observe a more restricted group of recombinants. The majority of these recombinants will have junctions lying within or close to regions A and B. Because class switch recombinants also include larger deletions which extend outside the switch regions, we can remove the kanamycin selection and look at a broader target zone. Both target zones are limited by the placement of the prokaryotic origin of replication downstream of region B. Resolution of events greater than 0.4 kilobases (kb) downstream of region B gives rise to molecules not detectable in the genetic assay. Transformation of switch sequence-bearing substrates such as pGD244 directly into *E. coli* without passing the plasmid through eukaryotic cells results in a background of white colonies of approximately 0.01% (10^{-4}) or lower.

3 Early Loss of Switch Sequence-Containing Substrates Following Transfection into Eukaryotic Cells

Upon transfection, we observe that a major fraction of the switch substrate DNA is lost compared with a cotransfected replicated molecule, pGD279 that carries the chloramphenicol marker. Loss occurs early, and it is important to note that it plateaus after 20 h in all cell lines examined (Fig. 2 and data not shown). Those substrate molecules which survive the 48-h transfection are intact, as determined by restriction enzyme digest and the ability of these molecules to undergo further recombination upon retransfection (data not shown). In addition, the retransfected substrate molecules manifest the same decrease in survival as the original substrate molecules (data not shown). Thus, the stabilization against plasmid loss established within 20 h after transfection does not appear to be genetic but epigenetic since it is reversible upon propagation in *E. coli*. Parallel experiments using nonswitch sequence substrate molecules, pGD209, showed no significant loss when compared with the second cotransfected molecule, pGD279 (Fig. 2). Recombination is measured as the ratio of recombinant to total plasmid molecules (the sum

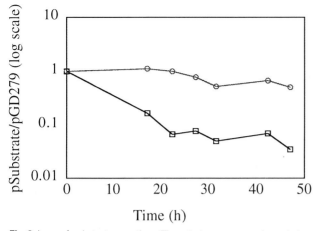

Fig. 2. Loss of substrate over time. The relative amounts of a switch sequence containing substrate, pGD244 (squares), and a nonswitch sequence containing substrate, pGD209 (circles), were compared after transfection into the pre-B line, 18–81. Both substrates were cotransfected with a second plasmid, pGD279, at a starting molar ratio of 1:1. The number of blue ampicillin-kanamycin resistant colonies from pGD244 was scored against the number of chloramphenicol resistant colonies from pGD279

of substrate and recombinant molecules). Reducing the number of substrates early after transfection could overestimate the percent recombination. In order to circumvent this potential source of background in the assay, we determine the rate of recombination during the interval from 24 to 48 h rather than relying on potentially inflated single time-point determinations at 48 h. The validity of this measurement will be considered in the following sections.

4 Recombination Is Switch Sequence and Time Dependent

Switch regions have been implicated as the *cis*-directing sequences responsible for class switching. We compared the recombination frequency of switch DNA (composed mostly of repetitive elements) with nonswitch DNA in an Abelson murine leukemia virus immortalized pre-B line, 18–81, and the mature B line, Bal17. Although class switch recombination normally is thought to be restricted to mature B cells in vivo, previous work has shown that some pre-B cells actively recombine their endogenous switch regions (Burrows et al. 1983; DePinho et al. 1984; Sugiyama et al. 1986; Yancopoulos et al. 1986). Recombination of our substrates is more than 50-fold greater for switch DNA than for similarly sized segments of either prokaryotic or eukaryotic DNA in both cell lines (Fig. 3). The slope is linear from 24 to 48 h and represents the rate of rec-

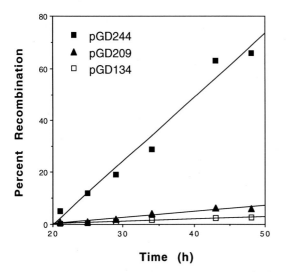

Fig. 3. Time course of recombination. Duplicate transfection points for three recombination substrates are plotted for the pre-B cell line 1–8 from 21 to 48 h after introduction into the cells. The broad target zone percent recombination determined by the percent of recombined ampicillin resistant recombinants is shown; the restricted target zone calculated from the percent of ampicillin (Ap) and kanamycin (Kn) resistant recombinant molecules has been published (DANIELS and LIEBER 1995). pGD134 has two different prokaryotic pieces of DNA inserted into regions A and B while pGD209 contains two different pieces of eukaryotic DNA fragments (DANIELS and LIEBER 1995). pGD244 contains Sμ in region A and Sγ3 in region B and is used as a positive control thoughout the study

ombination (calculated as the change in the percentage of recombined molecules per hour). Recombination measured selecting for recombinants on ampicillin (R-Ap), which is shown in Fig. 3) is always greater than the recombination level measured under ampicillin-kanamycin selection (R-ApKn) (not shown here but published in Fig. 2 of DANIELS and LIEBER 1995). The recombinants selected with ampicillin-kanamycin are a subset of recombinants whose junctions fall either in or immediately upstream (within 1 kb) of Sμ. Comparing the levels of recombination under the two selections gives an indication of the propensity for recombination to occur near Sμ or other sequences in region A on the substrate (Fig. 1).

5 Changes in Recombination Values 24 h After Transfection Are Primarily Due to De Novo Generation of Recombinant Molecules

A representative comparison of switch sequence substrate with control substrates at two time points illustrates that there is an absolute increase in recombinants over time (less than 2000 at 24 h up to over 20 000 at 48 h) (Table 1). The decrease in the total plasmid number (denominator) is illustrated in this data. This is typical of the decrease that occurs early after transfection, as described above.

To confirm that recombination values measured beginning 24 h after transfection reflect the generation of recombinants and not merely continued changes in substrate levels, two additional sources of background were addressed:
1. whether a substrate to product replication bias exists and
2. whether there is continued substrate loss.

Both phenomena would give rise to false recombination values over time. Although we measure recombination in the replicated pool, it is possible that replication may be slower for switch DNA substrates than for recombinant or nonswitch DNA substrates. To determine whether a replication bias exists, the rate of replication was measured for the switch and nonswitch DNA substrates. DNA initially transfected into the eukaryotic cells is marked with the prokaryotic *dam* methylation pattern. This methylation pattern is not maintained by the eukaryotic methylases and is lost during semiconservative DNA replication through a hemimethylated intermediate. By comparing the amount of hemimethylated DNA to fully demethylated DNA, one can determine the round of replication for a population of molecules. Such analysis shows no replication differences between switch DNA and nonswitch DNA containing substrates transfected into the pre-B cell lines, 1–8 and 18–81, or the mature B cell lines, M12 and Bal17 and measured out to 50 h (data not shown).

Replication bias is only one of many possibilities which might lead to preferential substrate loss over time. Therefore, we performed a second control

Table 1. Recombination of class switch and control substrates

Substrate	ApKn^r					R-ApKn (% change/h)
	24-h [white/blue + white]	%	48-h [white/blue + white]	%		
pGD134 (control)	0/85500	0	3000/67500	0.4		0.02
pGD209 (control)	150/93150	0.2	1200/571200	0.2		0.002
pGD244 (μ–γ3 switch substrate)	1950/36375	5.4	20400/48150	42.4		1.54

to address substrate loss in general. To accomplish this, the survival potential of switch substrates to recombinant molecules was compared. This comparison not only controls for any replication differences between the substrate and recombinant molecules, but also any other factor which might influence the number of plasmids (such as ongoing plasmid destruction due to non-productive recombination). However, the high rate of recombination makes direct determination of this value impossible. Instead we have again compared the survival of the switch substrate, pGD244, to a second plasmid, pGD279. An increase in the ratio of pGD279 to the sum of pGD279 and pGD244 represents preferential loss of pGD244 relative to pGD279. In essence, we are using pGD279 as a mock recombinant that can be observed independent of recombination.

The two substrates were cotransfected to give a ratio of 1:10 (pGD279:pGD244) 24 h post-transfection in the pre-B line 18–81. The change in the ratio of these two molecules was measured from 24 to 48 h post transfection in an analogous fashion as the rate of recombination (Table 2). Significantly, the relative level of the mock recombinant did not change from 24 to 48 h post transfection. In contrast, actual recombination of pGD244 (calculated from the same transfection) occurred at a high rate. The difference between the mock recombination and the real recombination is that the number of real recombinants can increase due to pGD244 product formation. Therefore, the rate of recombination measured after 24 h is not due to continued substrate loss but rather de novo generation of recombinants.

As a final control, a pool of 20 recombinants was cotransfected with pGD279 and found to be maintained at a constant ratio between 24 and 48 h post transfection (data not shown). Thus, pGD279 accurately represents the replication potential of recombinants compared to the substrate pGD244.

Table 2. Comparison of mock and actual recombination

Ratio	Recombination	Slope (Percent change in ratio per hour
pGD279/(pGD244 total + pGD279)	Mock	-0.054 ± 0.047
pGD244 recombinants/ pGD244 total	Actual	1.7 ± 0.6

6 Switch Recombination
Is Enhanced in the B Cell Lineage

It is not known which components of class switch recombination make the process cell lineage-specific and developmentally regulated. To begin to address this issue, the recombination activity of the substrates containing Sμ and Sγ3 was compared in a variety of hematopoietic cell lines (Fig. 4). Cell lines representing the mature B stage of development generally show the highest activity for recombination. In addition, both the pre-B cells and the plasmacytoma cell line, S194, manifest significant activity. On the other hand, all other cell types had relatively low recombination values. Most notably, six mature T lines had values lower than the lowest mature B cell line. The low level of recombination present in most cell lines is either an inherent background level of the switch substrates or illustrates a low level of switch recombination activity. It should be noted that compared with a nonswitch DNA-containing substrate, the switch DNA substrate gives a higher level of recombination in almost all cell lines (not shown). Clearly though, class switch recombination activity is greatest in the B cell lineage compared to other non-B cells. Our cell line survey results are consistent with a smaller survey by others (LEPSE et al. 1994). A potential second level of lineage and developmental stage specificity is discussed below.

7 Switch Regions Can Act Independently
in Recombination

The previous sections compared recombination of substrates containing either two switch regions or none. However, the question arises as to the recombination potential of one switch region alone. Evidence consistent with independent recombinational activation of switch regions has been reported (GU et al. 1993). The intra-switch sequence deletions, insertions, and inversions observed to occur in the genome (PETRINI and DUNNICK 1989) and the wide heterogeneity of length between inbred strains of mice suggest an instability associated with single switch regions (MARCU et al. 1980). Substrates were constructed containing either Sμ (pGD187) or Sγ3 alone (pGD259) combined with a segment of nonswitch DNA (Fig. 5). Recombination was measured in the pre-B line, 18–81, and the mature B line, Bal17.

Both switch regions independently show significant levels of recombination compared to the substrates containing nonswitch sequences, pGD134 and pGD209. In fact, the level of recombination when selecting on ampicillin-kanamycin for Sγ3 alone in the pre-B line 18–81 is only twofold lower than

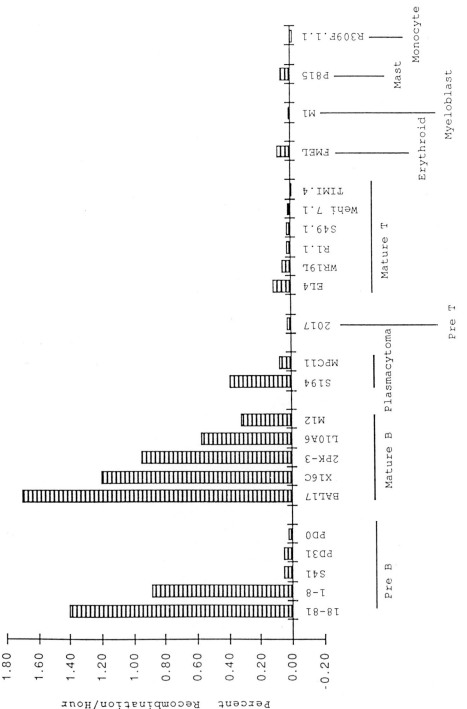

Fig. 4. Lineage and developmental stage predominance of class switch recombination activity

	Substrate	Cells	R-ApKn	R-Ap	Structure
1	pGD244	Bal17	1.7 ± 0.4	2.4 ± 0.7	μ / γ3
		18-81	1.4 ± 0.1	2.6 ± 0.3	
2	pGD134	Bal17	0.0054 ± 0.0025	0.023 ± 0.025	P1 / P2
		18-81	ND	ND	
3	pGD209	Bal17	-0.0052 ± 0.0016	0.050 ± 0.011	E2 / E1
		18-81	0.038 ± 0.017	0.058 ± 0.010	
4	pGD187	Bal17	0.15 ± 0.12	0.32 ± 0.17	μ / E1
		18-81	0.21 ± 0.04	0.85 ± 0.33	
5	pGD259	Bal17	0.59 ± 0.26	2.0 ± 0.1	E1 / γ3
		18-81	0.71 ± 0.07	2.2 ± 0.2	

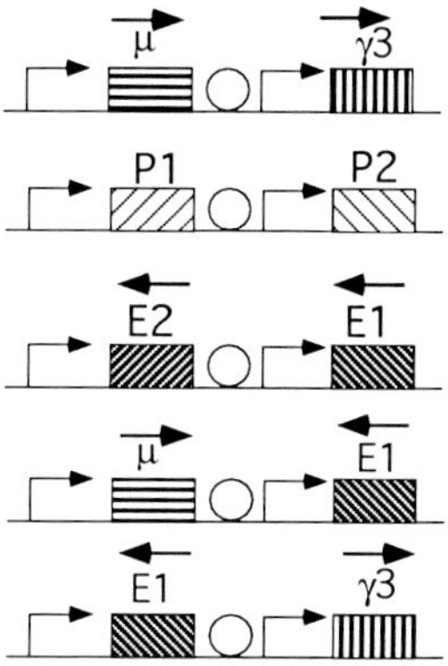

Fig. 5. Dependence of recombination on switch sequences. Bal17 is a mature B line, and 18–81 is a pre-B line (cf. Table 2). The plasmids pGD134 and pGD209 are nonswitch control DNA substrates. The substrate, pGD244, contains fragments of the murine Sμ and Sγ3. The plasmid, pGD187, contains a single switch region, Sμ, together with eukaryotic fragment E1. Likewise, pGD259 contains only Sγ3 together with the same eukaryotic fragment E1. The *arrows above the DNA* inserts illustrate the direction of transcription for these regions in the genome. The *circles* represent *supF*. The *arrows in front of each box* show the orientation of the eukaryotic promoters, SV/*tk* and hCMV (Fig. 1 and Sect. 2)

when present together with Sμ. However, in both cell lines, the level of re-combination is greater when both Sμ and Sγ3 are on the same molecule (pGD244) than simply adding the rates of recombination for the individual regions. In addition, the spectrum of recombination is altered to favor deletions more limited to the Sμ region (compare recombination rates between the ampicillin and the ampicillin-kanamycin selection values). Both observations argue that although single switch regions undergo independent recombination, the two switch regions together interact to reflect more than the sum of the individual switch regions.

8 Upstream Promoter Elements Regulate Class Switch Recombination

The heterologous promoters upstream of each switch region were removed in order to study their effects upon recombination. pGD255 is similar to pGD244 except for the removal of the SV/*tk* promoter in front of Sμ (Fig. 6). Removal of this promoter had no effect on the level of recombination. In contrast, a large effect is observed when the hCMV promoter upstream of Sγ3 is removed, pGD247. Recombination decreases 12-fold compared with pGD244. The drop is due to the effect of the hCMV promoter on Sγ3 and not due to the promoter

Substrate	R-ApKn	R-Ap	Structure
1 pGD244	1.7 ± 0.4	2.4 ± 0.7	
2 pGD189	-0.0035 ± 0.013	0.048 ± 0.020	
3 pGD247	0.14 ± 0.010	1.1 ± 0.59	
4 pGD255	1.8 ± 0.7	2.1 ± 0.41	
5 pGD275	-0.020 ± 0.009	0.037 ± 0.032	
6 pGD262	0.46 ± 0.040	0.58 ± 0.050	
7 pGD263	0.034 ± 0.035	0.087 ± 0.035	
8 pGD264	0.52 ± 0.15	0.56 ± 0.19	

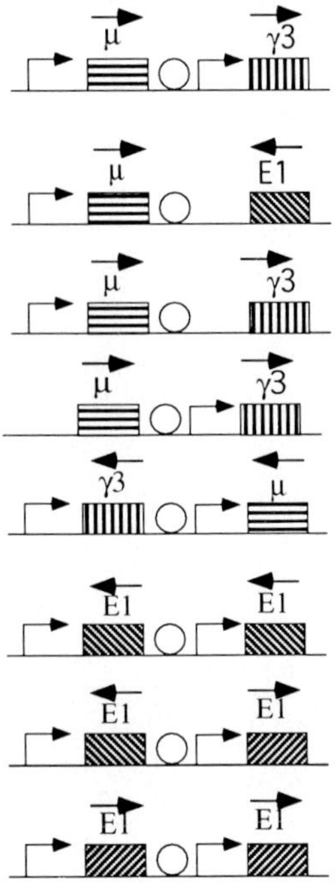

Fig. 6. Effect of promoters and orientation on switch-sequence recombination. The rate of recombination for substrates lacking promoters upstream of the switch regions was determined for the mature B cell line, Bal17. The fragments µ and γ3 are from switch regions while the fragment E1 is a similarly sized nonswitch DNA fragment. Substrates lacking promoters (pGD189, pGD247 and pGD255) are indicated by the corresponding promoter arrow being left off the figure. In *lines 6–8*, the rate of recombination was determined for the mature B cell line Bal17. For both pGD262 and pGD264, restriction analysis of the recombinants showed that greater than 90% were the result of homologous recombination between the two copies of E1 (not shown)

itself being a hotspot for recombination, (see pGD187 in Fig. 5). Using the substrate pGD189 as a baseline (background) for recombination of Sµ, both the hCMV promoter, pGD187, and the Sγ3, pGD247, increase recombination individually; however, together the promoter and the correctly oriented switch region generate the highest rate of recombination. Both Sγ3 and the hCMV promoter individually have values approximately one-tenth of when they are combined. Thus, hCMV appears to act in conjunction with Sγ3 to target recombination to this region. The lack of effect seen upon removal of the SV/*tk* promoter upstream of Sµ may be due to either a lower dependence of Sµ on transcriptional activation or an upstream promoter activity from either an adventitious promoter or read-through transcription from the polyoma large T gene.

9 Transcription Activates Recombination in a Strand-Specific Manner

The previous data indicate that recombination is dependent on switch region sequences and stimulated by transcriptional promoters. How might the promoters activate recombination? One possibility is a general opening of the chromatin to allow greater accessibility to the recombinase. Alternatively, transcription may promote the formation of a reaction intermediate by causing changes in DNA structure such as transient changes in supercoiling. A special case of this second possibility would be that the RNA molecule itself participates in maintaining this altered DNA structure. We tested a series of substrates to address these possibilities (Fig. 6). If transcription is required simply for increased chromatin accessibility to the recombinase, we might expect pGD275 to recombine at rates comparable to pGD244. Both substrates contain the same switch sequences and promoters but differ in the positioning of the fragments. The substrate pGD275 retains the original physiologic orientation of the switch regions to each other but changes transcription to the opposite strand.

The results are quite clear. The rates of recombination are substantially lower for pGD275 compared with pGD244 (Fig. 6). Because pGD275 retains the same promoters and orientation of the switch regions to one another, the low level of recombination eliminates "accessibility" as a possible role of transcription on our extrachromosomal substrates. The difference between pGD244 and pGD275 also points out that simple transcription of repetitive DNA is not sufficiently recombinogenic to account for the high rate of recombination seen in pGD244. Thus, switch regions have a defined strand specificity for transcriptional activation of recombination.

Although switch regions have low and variable degrees of homology to one another, we nonetheless wanted to formally test the possibility that homologous recombination may contribute to switch-sequence recombination. Though usually occurring by alternative splicing, DNA recombination from IgM to IgD can occasionally occur by homologous recombination involving non-switch region DNA (WHITE et al. 1990; YASUI et al. 1989). Sequence inspection of the more usual switch region-dependent recombination events shows no evidence for homologous recombination (DUNNICK et al. 1993; GERSTEIN et al. 1990; PETRINI and DUNNICK 1989; SZUREK et al. 1985). Nevertheless, the system here provides an opportunity for a direct test of this longstanding issue. Substrates pGD261 through pGD264 contain the same nonswitch DNA fragment in both regions A and B but in different orientations (Fig. 6). Homologous recombination is the basis for the high deletion rate between directly repeated sequences in substrates pGD262 and pGD264. However, it is important to note that there is no transcriptional strand bias for the recombination reaction. Substrates pGD262 and pGD264 are inverted with respect to transcription but both recombine well in either orientation. Hence, homologous recombination does not demonstrate sensitivity to transcription in one orientation versus the other. This is in striking contrast to the switch DNA substrate pair, pGD244 and pGD275. Thus, transcription in the physiologic orientation activates class switch recombination and uniquely distinguishes it from homologous recombination as well as other previously identified recombination pathways.

10 Discussion

For class switch recombination, the temporal correlation between activation of transcription and recombination is strong. The same exogenous factors which stimulate transcription from the upstream activating regions also direct class switching to that isotype (STAVNEZER et al. 1988). Sterile switch region transcription appears to be regulated both at the level of specific transcription factor production and chromatin accessibility around the sterile transcript promoter (XU and STAVNEZER 1992). However, whether this transcription is required for switch recombination has remained uncertain.

We constructed an extrachromosomal assay to study class switch recombination which allows the dissection of cellular and mechanistic requirements. In order to make inferences about the events at the endogenous loci, we show not only that recombination occurs on our substrates but that this recombination correlates in many ways to class switch recombination in the genome. As in the genome, recombination on our substrates is switch-sequence-dependent. Although the exactly analogous experiments employing similar size and base composition control DNA has never been performed in the genome, recombination on our substrates is over 300-fold greater for switch DNA than for nonhomologous, nonswitch DNA.

Secondly, endogenous class switch recombination is regulated in a tissue and developmentally specific manner. Recombination occurs predominantly at the mature B cell stage, but has also been found to occur earlier in some pre-B lines. Recombination in our assay is similarly restricted showing a strong predominance at the mature B cell stage as well as in some pre-B and plasmacytoma lines.

Thirdly, isotype switching is strongly associated with transcriptional activation of a targeted switch region. We find not just a correlation, but a dependence of recombination upon transcription of at least one of the switch regions. The transcription dependence is also orientation-specific. The transcriptional dependence of switch recombination that we have observed provides an additional level of lineage and developmental stage specificity because the sterile transcript promoters are only active in a lineage and developmental stage specific way (LUTZKER and ALT 1989; STAVNEZER-NORDGREN and SIRLIN 1986).

Finally, endogenous junctional endpoints involving Sμ are scattered over the entire Sμ region and occur to a significant extent both upstream and downstream of Sμ. In contrast, junctional endpoints for the acceptor switch region, γ3, fall almost exclusively in the repetitive region. The recombinants characterized using the extrachromosomal assay do not exclusively map to Sμ. In the pre-B cell line, 18–81, and in the mature B cell line, Bal17, approximately half of the recombinants lie in or within 1 kb upstream of Sμ. Of these, less than half appear to fall in the repetitive Sμ (data not shown). This is approximately twofold lower than what is seen in the genome (DUNNICK et al. 1993). On the other side of the junction, the majority of the sequenced recombinants (seven of eight) mapped to repetitive Sγ3 sequences. Several factors may contribute to the differences seen in our assay compared to the genome with respect to the breakpoint placement. First, our substrates use smaller switch regions which are positioned much closer to one another than their endogenous counterparts. Second, chromatin structure in the genome might influence the range of deletions. An interesting possibility is that we may be missing both sequences and/or proteins which define the recombination boundaries more precisely. This specificity, as noted, varies between cell lines and may represent the level of a factor dictating junctional position. Several proteins which might play such roles have been characterized (WUERFFEL et al. 1992; WILLIAMS and MAIZELS 1991; WATERS et al. 1989; LIAO et al. 1992).

Due to the nature of the assay, nonswitch DNA will always have a background level of recombination that is not present in the genome. Events leading to white colonies potentially can occur at four steps in our assay:

1. The propagation of substrates in *E. coli*,
2. The transfection of mammalian cells,
3. Propagation of substrates in mammalian cells, and
4. The isolation and measurement of recombinants in *E. coli*.

We can eliminate background problems at steps 1, 2 and 4 in our data by measuring the slope of recombination. The contribution of the *E. coli* background at both the initial substrate purification and at the isolation and measurement steps on the assay will be equivalent at 24 and 48 h post-transfection. Because we use the 24 h point as the initial value or background, the already low *E. coli* background present can be entirely factored out. Similarly, the background recombination rate due to the mutagenic effects of transfection can be eliminated in this fashion. Repair of damaged transfected molecules after 24 h to generate white colonies do not significantly contribute to the rate of recombination for two reasons. The first is that the majority of the repair processes appear to be essentially complete by 15 h, as measured by linear DNA transfection (G.A. DANIELS and M.R. LIEBER, unpublished results). Secondly, because the substrates are undergoing a high rate of replication, the number of repaired recombinant molecules will be very small compared with the larger substrate pool.

The level of background recombination due to mutations during replication in the mammalian cells (step 3 above) is not controlled for by determining the rate of recombination. To address this aspect of the background, similarly sized pieces of prokaryotic and eukaryotic DNA were studied as negative control substrates. Comparing the rate of recombination of these substrates with the level of recombination of the switch substrates gives the rate of switch sequence-dependent recombination. As seen in the results, this background level is very small (two to three orders of magnitude lower) compared with the recombination rate.

Recombination on our substrates differs from recombination of the endogenous locus in an interesting way. Recombination in the genome is regulated by upstream control regions. These upstream control regions are not included in our substrates. Instead, we utilized heterologous promoter elements upstream of each switch region. These promoters appear to contain the necessary features of the upstream control regions. In addition, the lack of higher order chromatin conformation of the substrates may eliminate the need for certain functions of the control region necessary for chromatin accessibility. In experiments not shown, cytokine responsiveness to interleukin (IL)-4 and interferon (IFN)-γ was tested in the cell lines 18–81 and Bal17. In all cases, no significant modulation of activity was measured in switching from Sμ to Sγ3 in our cellular assay. The nonresponsiveness is either the result of these cell lines lacking the necessary signaling machinery or that cytokine regulation occurs at the level of sterile transcription. Cytokine treatment is not expected to significantly modulate the level of these heterologous (hCMV and SV/tk) promoters. A definitive experiment to test the effects of cytokines using a population of primary splenic B cells for the transfection has not been possible for us, due to the difficulty in quantifying a system with a transfection efficiency that is only marginally above the background.

The data presented here is inconsistent with the regulation of recombination governed only at the level of the transcriptional enhancer and not the promoter itself (LEUNG and MAIZELS 1994). Our assay differs in several ways to the previously reported class switch episomal assay (LEUNG and MAIZELS 1992). Determination of the rate of recombination rather than reliance on single point measurements may be sufficient to explain the discordance between the two studies. The high background and substrate destruction outlined in our study shows that the rate of recombination rather than endpoint determinations more accurately reflect the recombination values. Although our results agree that a promoter may not be required upstream of Sμ, we find a dramatic stimulation of recombination when a promoter is placed upstream of Sγ3.

To summarize the levels of specificity of the recombination mediated by class switch sequences in this minichromosome substrate assay, we see over 300-fold specificity due to the switch sequences themselves (Fig. 5), 12-fold specificity due to transcription directed into the acceptor switch region (Fig. 6), and 20-fold specificity due to the cell lineage and developmental stage predominance of the activity that recombines the switch substrates. Combined, this is over 7500-fold specificity for recombination of transcribed switch sequences at the correct developmental stage within the B cell lineage.

The unique orientation dependence seen in our assay suggests a role for the promoter and transcription rather than an enhancer element or changes in supercoiling. Both the effects of an enhancer and changes in supercoiling would be expected to be orientation independent. Precisely how transcription directs recombination is unclear. It is intriguing to speculate that a similar RNA:DNA structure observed upon cell free transcription of Sα also directs the action of the recombinase (REABAN and GRIFFIN 1990). The same orientation of transcription which leads to a stable cell-free RNA:DNA association in Sα also leads to the activation of recombination. The evidence for a direct role for the sterile RNA transcript has remained tentative for lack of a functional assay system and the unique nature of the sequence of Sα for which the model was proposed. Extrapolating from the data presented here, we propose that the RNA molecule itself is necessary but not sufficient to direct recombination. Upon stimulation of a B cell to undergo class switch recombination, transcriptional induction may induce structural changes via the RNA transcript which target particular switch regions for recombination. Thus, locus targeting may, in part, be the direct result of transcriptional activation, and this would provide a level of tissue and stage-specificity due to the developmental regulation of sterile transcription of switch regions. This transcriptional targeting may be combined with the lineage and developmental stage predominance of class switch recombination activity to achieve a relatively exclusive recombination of switch regions observed in B lymphocytes.

Acknowledgments. The authors thank members of our laboratory for reading the manuscript. G.A.D. was supported by the PHS grant 5T32CA09302 awarded by the National Cancer Institute through the Program in Cancer Biology. This work was supported by NIH grants to M.R.L.

Note Added in Proof. Additional research on this topic from the author's laboratory can be found in "RNA:DNA Complex Formation Upon Transcription of Immunoglobulin Switch Regions: Implications for the Mechanism and Regulation of Class Switch Recombination." Nucl Acids Res 23 (24): 5006–5011 (1995) and in "Asymmetric Mutation Around the Breakpoint of Immunoglobulin Class Switch Sequences on Extrachromosomal Substrates." Nucl Acids Res 24 (11): (in press 1996).

References

Burrows PD, Beck-Engeser GB, Wabl MR (1983) Immunoglobulin heavy-chain class switching in a pre-B cell line is accompanied by DNA rearrangement. Nature 306:243–246

Clark DJ, Felsenfeld G (1992) A nucleosome core is transferred out of the path of a transcribing polymerase. Cell 71:11–22

Coffman RL, Lebman DA, Rothman P (1993) Mechanism and regulation of immunoglobulin isotype switching. Adv Immunol 54:229–270

Daniels GA, Lieber MR (1995) Strand-specificity in the transcriptional targeting of recombination at immunoglobulin class switch sequences. Proc Natl Acad Sci USA 92:5625–5629

Davis M, Kim SK, Hood LE (1980) DNA sequences mediating class switching in a-immunoglobulins. Science 209:1360–1365

Davis MM, Kim SK, Hood L (1980) Immunoglobulin class switching: developmentally regulated DNA rearrangements during differentiation. Cell 22:1–2

DePinho R, Kruger K, Andrews N, Lutzker S, Baltimore D, Alt FW (1984) Molecular basis of heavy-chain class switching and switch region deletion in an Abelson virus-transformed cell line. Mol Cell Biol 4:2905–2910

Dunnick WA, Hertz GZ, Scappino L, Gritzmacher C (1993) DNA sequence at immunoglobulin switch region recombination sites. Nucl Acid Res 21:365–372

Gerstein RM, Frankel WN, Hsieh C-L, Durdik JM, Rath S, Coffin JM, Nisonoff A, Selsing E (1990) Isotype switching of an immunoglobulin heavy chain transgene occurs by DNA recombination between different chromosomes. Cell 63:537–548

Gritzmacher CA (1989) Molecular aspects of heavy-chain class switching. Crit Rev Immunol 9:173–200

Gu H, Zou Y-R, Rajewsky K (1993) Independent control of immunoglobulin switch recombination at individual switch regions evidenced through Cre-loxP-mediated gene targeting. Cell 73:1155–1164

Hesse JE, Lieber ML, Gellert M, Mizuuchi K (1987) Extrachromosomal substrates undergo inversion or deletion at immunoglobulin VDJ joining signals. Cell 49:775–783

Jack HM, McDowell M, Steinberg CM, Wabl M (1988) Looping out and deletion mechanism for the immunoglobulin heavy-chain class switch. Proc Natl Acad Sci 85:1581–1585

Jung S, Rajewsky K, Radbruch A (1993) Shutdown of class switch recombination by deletion of a switch region control element. Science 259:984–987

Lepse CL, Kumar R, Ganea D (1994) Extrachromosomal eukaryotic DNA substrates for switch recombination: analysis of isotype and cell specificity DNA. Cell Biol 13:1151–61

Leung H, Maizels N (1992) Transcriptional regulatory elements stimulate recombination in extrachromosomal substrates carrying immunoglobulin switch-region sequences. Proc Natl Acad Sci 89:4154–4158

Leung H, Maizels N (1994) Regulation and targeting of recombination in extrachromosomal substrates carrying immunoglobulin switch region sequences. Mol Cell Biol 14:1450–1458

Liao F, Giannini SL, Birshtein BK (1992) A nuclear DNA-binding protein expressed during early stages of B cell differentiation interacts with diverse segments within and 3' of the Ig H chain gene cluster J Immunol 148:2909–2917

Lieber ML, Hesse JE, Mizuuchi K, Gellert M (1987) Developmental stage specificity of the lymphoid V(D)J recombination activity. Genes Dev 1:751–761

Lutzker SG, Alt FW (1989) Immunoglobulin heavy-chain class switching. In: Berg DE, Howe MM (eds) Mobile DNA. American Society for Microbiology, Washington, DC, pp 693–714

Marcu KB, Banerji J, Penncavage NA, Lang R, Arnheim N (1980) 5' flanking region of immunoglobulin heavy chain constant region genes display length heterogeneity in germlines of inbred mouse strains. Cell 22:187–196

Ott DE, Alt FW, Marcu KB (1987) Immunoglobulin heavy chain switch region recombination within a retroviral vector in murine pre-B cells. EMBO 6:577–584

Petrini J, Dunnick W (1989) Products and implied mechanism of H chain switch recombination. J Immunol 142:2932–2935

Reaban ME, Griffin JA (1990) Induction of RNA-stabilized DNA conformers by transcription of an immunoglobulin switch region. Nature 348:342–344

Schwedler UV, Jack H-M, Wabl M (1990) Circular DNA is a product of the immunoglobulin class switch rearrangement. Nature 345:452–456

Stavnezer J, Abbott J, Sirlin S (1984) Immunoglobulin heavy chain switching in culture I29 murine B lymphoma cells:commitment to an IgA or IgE switch. Curr Topics Microbiol Immunol 113:109–115

Stavnezer J, Radcliff G, Lin Y-C, Nietupski J, Berggren L, Sitia R, Severinson E (1988) Immunoglobulin heavy-chain switching may be directed by prior induction of transcripts from constant-region genes. Proc Natl Acad Sci USA 85:7704–7708

Stavnezer-Nordgren J, Sirlin S (1986) Specificity of immunoglobulin heavy chain switch correlates with activity of germline heavy chain genes prior to switching EMBO 5:95–102

Sugiyama H, Maeda T, Akira S, Kishimoto S (1986) Class-switching from μ to γ3 or γ2b production at pre-B cell stage. J Immunol 136:3092–3097

Szurek P, Petrini J, Dunnick W (1985) Complete nucleotide sequence of the murine γ3 switch region and analysis of switch recombination sites in two γ3-expressing hybridomas. J Immunol 135:620–626

Waters SH, Saikh KU, Stavnezer J (1989) A B-cell-specific nuclear protein that binds to DNA sites 5' to immunoglobulin Sα tandem repeats is regulated during differentiation. Mol Cell Biol 9:5594–5601

White MB, Word CJ, Humphries CG, Blattner FR, Tucker PW (1990) Immunoglobulin D switching can occur through homologous recombination in human B cells. Mol Cell Biol 10:3690–3699

Williams M, Maizels N (1991) LR1, a lipopolysacharide-responsive factor with binding sites in the immunoglobulin switch regions and heavy-chain enhancer. Genes Dev 5:2353–2361

Wuerffel R, Jamieson CE, Morgan L, Merkulov GV, Sen R, Kenter AL (1992) Switch recombination breakpoints are strictly correlated with DNA recognition motifs for immunoglobulin Sγ3 DNA-binding proteins J Exp Med 176:339–349

Xu L, Gorham B, Li SC, Bottaro A, Alt FA, Rothman P (1993) Replacement of germ-line e promoter by gene targeting alters control of immunoglobulin heavy chain class switching. Proc Natl Acad Sci 90:3705–3709

Xu M, Stavnezer J (1992) Regulation of transcription of immunoglobulin germ-line g1 RNA: analysis of the promoter/enhancer. EMBO 11:145–155

Yancopoulos GD, DePinho RA, Zimmerman KA, Lutzker SG, Rosenberg N, Alt FW (1986) Secondary genomic rearrangement events in pre-B cells: VhDJh replacement by a LINE-1 sequence and directed class switching. EMBO 5:3259–3266

Yasui H, Akahori Y, Hirano M, Yamada K, Kurosawa Y (1989) Class switch from μ to δ is mediated by homologous recombination between σμ and Σμ sequences in human immunoglobulin gene loci. Eur J Immunol 19:1399–1403

Zhang J, Bottaro A, Li S, Stewart V, Alt F (1993) A selective defect in IgG2b switching as a result of targeted mutation of the Iγ2b promoter and exon. EMBO J 12:3529–3537

Biochemical Studies
of Class Switch Recombination

Rolf Jessberger[1], Matthias Wabl[2], and Tilman Borggrefe[1]

1 Introduction

Higher eukaryotes produce immunoglobulins (Ig) of different classes, which are defined by the constant region (C) of the heavy chain. In the mouse the heavy chains are designated μ, δ, $\gamma3$, $\gamma1$, $\gamma2b$, $\gamma2a$, ϵ, and α. Upon stimulation by antigen, expression of the early IgM class changes to that of another class (IgG3, IgG1, IgG2b, IgG2a, IgE, or IgA) can occur in B lymphocytes. Because the variable region is retained during this switch, the antigen specificity is unaltered. At the DNA level the process, named simply "class switching" from hereon, occurs through a DNA recombination event (for reviews see Esser and Radbruch 1990; Harriman et al. 1993).

Apart from Cδ, all the constant region genes are preceded by a switch (S) region of several kilobasepairs (kbp) in length (for review see Shimuzu and Honjo 1984). They contain many short, G rich repetitive sequence elements, often arranged in tandem arrays. Although there is no consensus recombination signal sequence comparable to the heptamer-nonamer sequence for V(D)J recombination, the pentameric motifs GAGCT, GGGGT, and GGTGG, present in all S regions, were found often at or near the recombination sites (Petrini and Dunnick 1989). The unstable, recombinogenic nature of S regions can be seen in cells, which have not yet switched, but have deletions in Sμ (Winter et al. 1987).

There is a preference for class switch recombination to take place within the S regions, especially in the pentameric repeat regions (Gritzmacher 1989).

---

[1] Basel Institute for Immunology, Grenzacherstrasse 487, 4005 Basel, Switzerland
[2] Department of Microbiology and Immunology, University of California, San Francisco, San Francisco, CA 94143–0670, USA

In marked contrast to those observed in V(D)J rearrangements (for reviews, see GELLERT 1992; LIEBER 1992; SCHATZ et al. 1992; JESSBERGER 1994; and D.C. VAN GENT et al., this volume), the recombination break-and-rejoining points are not precise: they are scattered over the S regions and, in some instances, in close proximity to but outside of the S regions. Therefore, class switch recombination may be more accurately described as region-specific rather than site-specific recombination. Thus, S regions greatly differ from the V(D)J recombination signal sequences in that they are relatively heterogeneous, extended, and unprecise in guiding to the breakpoint. In some instances, DNA sequences derived from three S regions have been detected after rearrangement, indicating multiple switch recombination events (PETRINI and DUNNICK 1989), e.g., from IgM to IgG1 and then to IgE. Thus, a once recombined S region remains accessible for secondary events.

Sequence homologies between S regions vary; for example, when compared to human Sμ the human Sγ4 region displays the lowest and the Sε region the highest degree of homology (MILLS et al. 1990). The switch from IgM to IgG, however, is the most frequent one, rendering a simple relationship between the extent of homology and frequency of recombination between S regions unlikely. On the other hand it is still possible that short patches of homology between the S regions can be used as nucleation points by a recombination machinery to facilitate formation of S region interactions, formation of ternary complexes and initiation of the rearrangement. The use of short patches of homology has been demonstrated in mammalian cells for homologous (ANDERSON et al. 1984) and nonhomologous recombination (for reviews see ROTH and WILSON 1988; DOERFLER et al. 1983, 1989).

The recombinative mechanism for most class switching events can be described by the loop-excision model (JÄCK et al 1988; Fig. 1). Following transcriptional activation (discussed by M. Lorenz and A. Radbruch, this volume), which is induced by antigenic stimulation of the B lymphocytes, DNA recombination deletes Cμ and Cδ, transferring the γ, α, or ε constant region genes closer to the rearranged and expressed VDJ gene segments. The model postulates two steps, loop formation and excision. The deleted DNA fragment exists in the B cell for a while as circular DNA of various sizes, depending which particular switch region was involved in the recombination event (VON SCHWEDLER et al. 1990; IWASATO et al. 1990). The loop and excision model involves the creation of four free DNA ends after endonucleolytic cleavage of the two S regions to be recombined. These DNA ends are possibly protected or even fixed by hitherto unknown proteins. The exact structure of the loop intermediate is not known, and may resemble a crossed (α), a stem-loop (Ω), or another conformation. The four ends can be rejoined in their original configuration, joined in a way to generate an inversion, or the intervening sequence can be circularized and removed while the remaining chromosomal DNA ends are joined together to yield the switched gene (JÄCK et al. 1988). Both inversions and excised circular DNAs, the so-called switch circles, have been observed in switching cells, providing evidence for the model (VON SCHWEDLER et al. 1990).

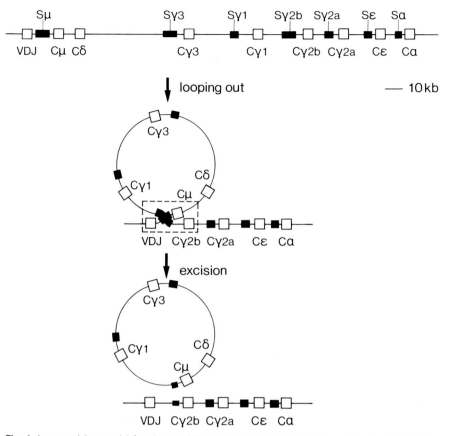

Fig. 1. Loop-excision model for class switch recombination. Modified from Jäck et al. 1988. The *stippled box* illustrates a hypothetical protein-DNA complex

2 The Putative Switch Recombinase

Considering the loop-excision model shown in Fig. 1, a minimum of enzymatic activities in class switch recombination should include: 1. a factor that recognizes S regions, 2. an endonuclease that cleaves within the S regions, and 3. a DNA joining activity, e.g., a ligase, which seals the DNA ends generated by the nuclease. Additional proteins might be necessary, e.g., to facilitate the formation of a hypothetical complex formed by the two S regions. This assumes that the S regions, which can be located more than 100 kbp apart on the chromosome, are brought together before endonucleolytic cleavage of an S region. Alternatively, it is conceivable that the S regions are brought together only after one of them has been cleaved, or that both S regions are cleaved without forming a close complex. Then for the chromosomal ends to be rejoined, they would be required to move towards each other. This also requires that standard DNA double-strand break repair mechanisms do not operate during switching. They would join the ends with any other, nonS region end or would initiate homologous recombinational repair with the intact allele as the template. To date, neither has the endonuclease been characterized, nor is it known what type of ends are being generated by S region cleavage, e.g., whether these ends are phosphorylated at their 5' termini and whether they are ligatable. In V(D)J recombination, hairpins are formed at the coding ends (see S. BOCKHEIM STEEN et al., this volume). Due to the very imprecision of class switch recombination sites, it seems unlikely that similar intermediate structures are generated during that reaction. However, DNA end modifying enzymes, e.g., DNA polymerases or exonucleases, might well be processing the ends prior to joining. Even if these modifiers were not necessary, they might nevertheless act on the ends and/or compete for them with the joining activity. Indeed, frequent mutations like base substitutions, insertions, and deletions at switch recombination junctions (WINTER et al. 1987; DUNNICK et al. 1989), suggest the involvement of DNA metabolic enzymes, e.g., a polymerase, in the reaction. Joining itself may proceed by topoisomerase- or ligase-like mechanisms.

Finally, it is possible that two or more of the enzymatic functions mentioned above reside within one protein.

Unlike V(D)J recombination, where RAG-1 and RAG-2 are needed specifically (see D.C. VAN GENT et al., and D.G. SCHATZ and T.M.J. LEU, this volume), no such proteins have been described for class switching. The RAG proteins are not necessary for class switching (A. ROLINK, F. MELCHERS, and J. ANDERSSON, unpublished results). This again stresses the difference of the two recombination reactions necessary for the function of the immune system.

Certainly one of the most compelling questions is that of specificity: Is there a recombination machinery that exclusively acts on S regions, or is there a general machinery which is directed to the S regions by one or more specificity factors? Or is there no specific recombination factor at all, and S region recombination is stimulated by transcriptional opening of the locus, and/or the highly repetitive structures and the G/C richness of the S regions guiding recombination? In this scenario, the S regions could be viewed as hot spots for recombination in a general, not in a heavy chain gene-specific sense, just as there are other hot spots in the genome. Specificity would be achieved by other factors, e.g., those associated with transcriptional activators of the regions. The so-called germ line transcripts, which start 5' of the S regions and terminate 3' of the respective C_H genes, may be directly involved in determining the specificity (see M. Lorenz and A. Radbruch, this volume). Today, it seems impossible to decide between these possibilities. What has also hampered research is the fact that, in contrast to V(D)J recombination, there are no mutants analogous to the *scid* (severe combined immunodeficiency) mouse (BOSMA and CARROLL 1991), or the RAG-1 and -2 knockout mice (MOMBAERTS et al. 1992; SHINKAI et al. 1992). The mutants with differences in class switching such as *xid* have a defect in signalling rather than in the proteins directly involved in recombination.

3 Switch Region Specific DNA Binding Proteins

Since class switch recombination is dependent on S regions with their particular characteristics, a number of groups searched for proteins that specifically bind to the S regions. A predominant technique used in these investigations is the gel mobility shift assay, which may reveal specificity, but which allows only very limited conclusions about possible enzymatic or other functions of the proteins detected by the assay.

An octamer-like sequence, ATGCAAAA, which matches seven of eight nucleotides of the transcriptional enhancer motif, has been detected by DNAseI hypersensitivity assays in the Sγ1 region, and nuclear proteins binding specific to this site and its flanking sequences was observed (SCHULTZ et al. 1991). However, the proteins binding in gel shift experiments to this sequence were derived from a B cell hybridoma line and could not be distinguished from known octamer-binding factors OTF-1 and OTF-2. Another protein, called Sα-BP, which is specifically expressed in pre-B and B lymphocytes and not found in other B or T lineage derived cells or in nonlymphoid cells, binds to sites 5' of the Sα tandem repeats (WATERS et al. 1989), but not to the repeats themselves. A protein, binding to the Sμ region, NF-Sμ, was identified from lipopolysaccharide (LPS)- or dextran sulfate-stimulated mouse splenic B cells (WUERFFEL et al. 1990). It binds most efficiently to a 25 bp Sμ repetitive sequence element, of which a stretch of four G residues flanked by GAGCT was found to be optimal for binding. Finally, LR1, a 106 kDa protein responsive to LPS treatment of splenic B cells was found to bind to switch regions as well as heavy chain gene enhancers (WILLIAMS and MAIZELS 1991). The protein recognizes a G-rich sequence that is frequent in switch regions. It seems to bend DNA and to form higher order complexes, probably through multimeric binding. LR1 has now been purified from murine preB cells; it is phosphorylated and N-acetylglucosaminylated (WILLIAMS et al. 1993). It turned out to be a transcription factor regulating expression of several genes including c-*myc*, whose promotor/enhancer region also contains G-rich regions. It has been hypothesized that LR1 may be involved in class switch recombination as well as in c-*myc* translocation and transcription.

None of these binding proteins, however, has directly been shown to be involved in class switch recombination, e.g., by genetic knockout or biochemical experiments, and their role in recombination – if any – remains to be determined.

Specific binding to S regions may be an important step during initiation of class switching. The first enzymatic step, however, most likely would be the incision made by an endonuclease. This endonuclease may be a general one, recruited to the S regions by a specificity factor, or may itself be specific for S regions. The search for such an endonuclease, which should be active in switching B cells – but not necessarily expressed exclusively in these cells – is underway in several laboratories, but so far no reports have been published.

Another strategy, sometimes dubbed "in vivo biochemistry", which makes use of episomal DNA substrates containing S regions, transfected into a variety mammalian cells, is described in the chapter by G.A. Daniels and M.R. Lieber, this volume.

4 Towards a Cell-Free Switch Region Specific Recombination Reaction

A different approach to study the enzymes possibly involved in class switching was taken in the authors' laboratory. Based on the loop-excisison model shown in Fig. 1, we devised an assay which makes use of two S (Sµ and Sγ2b) regions, cloned into two different, largely nonhomologous vectors (Fig. 2). The method described in the following is a variation of the DNA Transfer Assay (DTA) (JESSBERGER and BERG 1991) which has been applied in studies of mammalian homologous recombination activites (JESSBERGER et al. 1993) and mammalian recombination mutants (JESSBERGER et al. 1995). One of the two partners to be used as recombination substrates was prepared as plasmid DNA, uniformly in vivo labeled with [3H]thymidine. The other was modified to contain a small number of digoxigenin (dig) molecules.

Upon incubation in nuclear extracts derived from induced B cells which undergo Ig class switching, one would expect among the possible recombination products those which are depicted in Fig. 2, where the tritiated S region plasmid had stably recombined with the dig-S-substrate. Binding the products to anti-digoxigenin antibody beads and measuring the retained [3H] radioactivity gives a direct measurement of DNA transfer between the two S regions. This assay is quick and quantitative and therefore suitable for protein purification experiments. For controls, DNA substrates lacking the S regions are used. The design of the assay restricts the reaction to the intermolecular type, which in a way resembles the reverse loop-excision reaction – a reintegration pathway. Intermolecular class switching between different chromosomes has been observed in vivo (KNIGHT et al. 1974; KIPPS and HERZENBERG 1986; GERSTEIN et al. 1990). The crucial point, however, is the dependence of the activity on the presence of the two S regions.

The overall experimental design (Fig. 3) includes the preparation of splenic B cells from mice, their stimulation to switch by treating them with LPS, and subsequently the preparation of nuclear extracts from the switch-induced cells. Nuclei were extracted by salt treatment according to published procedures

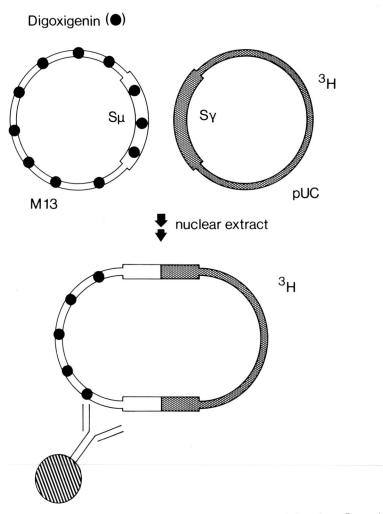

Fig. 2. Biochemical assay to detect DNA transfer between switch regions. For explanations, see text

(JESSBERGER et al. 1993). Other controls included the omission of LPS and preparation of extracts from nonswitching cells or the separation of switching from nonswitching cells by cell elutriation methods and subsequent extract preparation from both populations.

In the crude nuclear extracts prepared from the total, LPS-treated spleen cell culture, a variety of recombination activities are present that act on homologous and heterologous DNA substrates, which possibly mask S region specific activity. Therefore, the extracts were immediately fractionated on one or two chromatography columns (unpublished results). When we assayed the protein fractions with DNA substrates containing (or, as a control, lacking) the

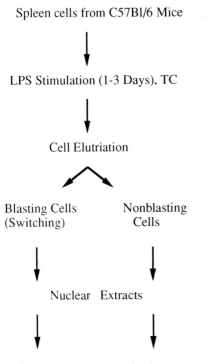

Spleen cells from C57Bl/6 Mice

LPS Stimulation (1-3 Days), TC

Cell Elutriation

Blasting Cells   Nonblasting
(Switching)   Cells

Nuclear   Extracts

Cell Free DNA Transfer Assay

Fig. 3. Experimental design for a cell-free method to study recombination between switch regions (*TC*, tissue culture; *LPS*, lipopolysaccharide)

two S regions we found an activity that preferentially recombines the S region substrates (Fig. 4). The preference increases from about twofold in protein fractions from purification step II (fraction II) to approximately tenfold in fraction III. This activity has not been found in noninduced splenic B cell extract preparations. The cytoplasmic fractions had no activity. Separating switching from nonswitching cells by elutriation prior to extract preparation resulted in an enrichment of S region specific activity. By doing so we observed a twofold preference for S region substrates already in the crude nuclear extracts.

Results from incubation of the resulting active fractions with various DNA substrates are given in Table 1. In summary the DNA transfer activity is highest when the two S region substrates were used. Reactions with only one S region substrate are less than half as efficient as with two S substrates. Replacement of both S substrates by the respective vector DNAs yielded less than 20% activity as compared to the reaction with the two S region substrates. Also, homologous DNA substrates – derivatives of the pSV2neo plasmid, which have been widely used in similar experiments before (JESSBERGER and BERG 1991; JESSBERGER et al. 1993, 1995) – served as poor substrates for fraction II and were only half efficient substrates for fraction III. As discussed above, a role of short patches of homology in class switch recombination cannot be

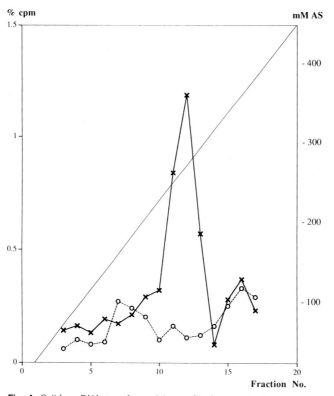

Fig. 4. Cell-free DNA transfer activity profile for protein fractions at step III of purification. DNA substrates with 2 S regions (*solid line*) are compared for each protein fraction with substrates lacking S regions (*stippled line*; *AS*, ammonium sulfate;% *cpm*, the portion of input radioactivity of one substrate having recombined with the other; *1% cpm ≙ 1400 cpm*)

neglected, and it might well be that parts of the S region preferring recombination activity act more or less efficiently on homologous substrates as well.

These data allow experiments to determine the precise structure of the recombination products obtained in this scheme and to further purify the proteins responsible for the activity. The approach may or may not result in the isolation of the "true" switch recombinase, i.e., the enzyme complex, which performs switch recombination in the induced B lymphocytes. The experiments, however, will at the minimum yield a new and interesting DNA transfer activity, which seems to be specific for switch-induced B cells.

Table 1. Cell-free DNA transfer of switch region substrates

DNA substrates	Protein fraction	Activity (%)
2 S Regions (Sμ-M13 + Sγ-pUC)	II, 300 ng	100[c]
1 S Region (M13 + Sγ-pUC)	II, 300 ng	28
1 S Region (Sμ-M13 + pUC)	II, 300 ng	46
No S Region (M13 + pUC)	II, 300 ng	11
No S Region, full homology[a]	II, 300 ng	15
2 S Regions (Sμ-M13 + Sγ-pUC)	III, 2 ng	100[d]
1 S Region (M13 + Sγ-pUC)	III, 2 ng	44
1 S Region (Sμ-M13 + pUC)	III, 2 ng	25
No S Region (M13 + pUC)	III, 2 ng	18
No S Region, full homology[a]	III, 2 ng	60
No S Region, partial homology[b]	III, 2 ng	60
2 S Regions (Sμ-M13 + Sγ-pUC)	III, 2 ng, inactivated	10

[a]pSV2neoΔ70 + pSV2neo;
[b]pSV2neoΔ70 + pUC;
[c]100% = 1700 cpm;
[d]100% = 1200 cpm

5 Conclusions

One of the most fascinating quests in the molecular analysis of class switching is certainly the search for the factor or reaction step which brings about the specificity for S regions within the IgH locus. As discussed in Sect. 2, specificity may be determined on a variety of levels, either at one or more steps. These include specific transcription of the germ line DNA over the target S region, where either protein components of the transcription apparatus or the RNA itself could provide specificity. It has also been suggested, that the methylation status of the DNA and a hypothetical (de)methylation system may contribute to specificity (BURGER and RADBRUCH 1990; KOCHANEK et al. 1991). In any of these cases, however, one could view the respective factors as components of the switch recombination machinery as well. The unique features of the S regions make it likely that specificity is at least codetermined by these remarkable sequences. Due to the size and heterogeneity of S regions and the distribution of recombination sites, a simple sequence specificity seems unlikely. However, recognition of secondary structures, G richness or repetitive motifs, like the pentamers in the S regions, may provide a means for specificity.

This in turn raises the question for proteins or protein complexes which specifically bind to S regions and act on them as outlined above. Proteins identified in gel mobility shift assays that bind to S regions are candidates. The evidence for their direct involvement in the recombination reaction however is still missing. Since these proteins should predominantly act at the initiation steps of class switch recombination, other approaches might be necessary to obtain insights into the complete reaction. One such approach is outlined in Sect. 4, where final products of a cell-free switch region dependent recombination reaction can be looked at.

In addition to the search for enzymatic activities performing class switch recombination, many questions remain to be answered, especially about molecular and mechanistic aspects of the reaction: Are both S regions recognized simultaneously, even though they may be more than100 kbp apart from each other? Alternatively, does one S region serve as loading template for an enzyme complex sliding then along the DNA in search for the second accessible S region? How does the chromatin structure of the DNA influence the reaction and must the DNA be bend? What is the structure of the hypothetical complex intermediate containing the two S regions and the enzymes? Is there any direct role for transcription or components of the transcription apparatus in the recombination reaction? Is rejoining of switch circle ends coupled to rejoining of the chromosomal DNA ends? At least some of these problems should become accessible to direct studies as suitable cell-free model systems for class switching are now being developed.

Acknowledgements. The authors thank Dr. Jan Andersson for critical reading of the manuscript. The Basel Institute is founded and supported by Hoffmann-LaRoche Ltd., Switzerland. Work of M.W. is supported by NIH grant 1R01 GM37699 and by funds from the Markey Trust.

References

Anderson RA, Kato S, Camerini-Otero RD (1984) A pattern of partially homologous recombination in mouse L cells. Proc Natl Acad Sci USA 81:206–210

Bosma MJ, Carroll AM (1991) The SCID mouse mutant: definition, characterization, and potential uses. Annu Rev Immunol 9:323–350

Burger C, Radbruch A (1990) Protective methylation of immunoglobulin and T cell receptor (TcR) gene loci prior to induction of class switch and TcR recombination. Eur J Immunol 20:2285–2291

Doerfler W, Gahlmann R, Stabel S, Deuring R, Lichtenberg U, Schulz M, Eick D, Leisten R (1983) On the mechanism of recombination between adenoviral and cellular DNAs:the structure of junction sites. Curr Top Microbiol Immunol 109:193–228

Doerfler W, Jessberger R, Lichtenberg U (1989) Recombination between adenovirus DNA and the mammalian genome. Curr Top Microbiol Immunol 144:209–216

Dunnick W, Wilson M, Stavnezer J (1989) Mutations, duplications, and deletion of recombined switch regions suggest a role for DNA reolication in the immunoglobulin heavy chain switch. Mol Cell Biol 9:1850–1856

Esser C, Radbruch A (1990) Immunoglobulin class switching:molecular and cellular analysis. Annu Rev Immunol 8:717–735

Gellert M (1992) Molecular analysis of V(D)J recombination. Annu Rev Genet 22:425–446

Gerstein RM, Frankel WN, Hsieh C-L, Durdik JM, Rath S, Coffin JM, Nisonoff A, Selsing E (1990) Isotype switching of an immunoglobulin heavy chain transgene occurs by DNA recombination between different chromosomes. Cell 63:537–548

Gritzmacher CA (1989) Molecular aspects of heavy-chain class switching. Crit Rev Immunol 9(3):173–200

Harriman W, Völk H, Defranoux N, Wabl M (1993) Immunoglobulin class switch recombination. Annu Rev Immunol 11:361–384

Iwasato T, Shimizu A, Honjo T, Yamagishi H (1990) Circular DNA is excised by immunoglobulin class switch recombination. Cell 62:143–149

Jäck H-M, McDowell M, Steinberg C M, Wabl M (1988) Looping out and deletion mechanism for the immunoglobulin heavy-chain class switch. Proc Natl Acad Sci USA 85:1581–1585

Jessberger R (1994) Biochemical aspects of DNA recombination in the immune system. The Immunologist 2:201–207

Jessberger R, Berg P (1991) Repair of deletions and double-strand gaps by homologous recombination in a mammalian in vitro system. Mol Cell Biol 11:445–457

Jessberger R, Podust V, Hbscher U, Berg P (1993) A mammalian protein complex that repairs double-strand breaks and deletions by recombination. J Biol Chem 268:15070–15079

Jessberger R, Riwar B, Rolink A, Rodewald HR (1995) Stimulation of defective DNA transfer activity in recombination deficient scid cell extracts by a 72 kDa protein from wildtype thymocytes. J Biol Chem 270:6788–6797

Kipps TJ, Herzenberg LA (1986) Homologous chromosome recombination generating immunoglobulin allotype and isotype switch variants. EMBO J. 5:263–268

Knight KL, Malek TR, Hanly WC (1974) Recombinant rabbit secretory immunoglobulin molecules:alpha chains with matural (paternal) varibale-region allotypes and paternal (maternal) constant-region allotypes. Proc Natl Acad Sci USA 71:1169–1173

Kochanek S, Radbruch A, Tesch H, Renz D, Doerfler W (1991) DNA methylation profiles in the human genes for tumor necrosis factors a and b in subpopulations of leukocytes and in leukemias. Proc Natl Acad Sci USA 88:5759–5763

Lieber MR (1992) The mechanism of V(D)J recombination:a balance of diversity, specificity and stability. Cell 70:873–876

Mills FC, Brooker JS, Camerini-Otero RD (1990) Sequences of human immunoglobulin switch regions: implications for recombination and transcription. Nucleic Acids Res 18:7305–7316

Mombaerts P, Iacomini J, Johnson RS, Herrup K, Tonegawa S, Papaiannou VE (1992) RAG-1 deficient mice have no mature B and T lymphocytes. Cell 68:869–877

Petrini J, Dunnick WA (1989) Products and mechanism of H chain switch recombination. J Immunol 142:2932–2935

Roth D, Wilson J (1988) Illegetimate recombination in mammalian cells. In: Kucherlapati R, Smith GR (eds) Genetic recombination. ASM, Washington DC, pp 621–654

Schatz DG, Oettinger MA, Schlissel MS (1992) V(D)J recombination:molecular biology and regulation. Annu Rev Immunol 10:359–383

Schultz CL, Elenich LA, Dunnick WA (1991) Nuclear protein binding to octamer motifs in the immunoglobulin g1 switch region. Intern Immunol 3:109–116

Shimuzu A, Honjo T (1984) Immunoglobulin class switching. Cell 36:801–803

Shinkai Y, Rathburn G, Lam KP, Oltz EM, Stewart V, Mendelsohn M, Charron J, Datta M, Young F, Stall AM, Alt FW (1992) RAG-2 deficient mice lack mature lymphocytes owing to inability to initiate V(D)J recombination. Cell 68:855–867

von Schwedler U, Jäck HM, Wabl M (1990) Circular DNA is a product of the immunoglobulin class switch rearrangement. Nature 345:452–454

Waters SH, Saikh KU, Stavnezer J (1989) A B-cell-specific nuclear protein that binds to DNA sites 5' to immunoglobulin Sα tandem repeats is regulated during differentiation. Mol Cell Biol 9:5594–5601

Williams M, Maizels N (1991) LR1, a lipopolysaccharide-responsive factor with binding sited in the immunoglobulin switch regions and heavy chain enhancer. Genes Dev 5:2353–2361

Williams M, Hanakahi LA, Maizels N (1993) Purification and properties of LR1, an inducible DNA binding protein from mammalian B lymphocytes. J Biol Chem 268:13731–13737

Winter E, Krawinkel U, Radbruch A (1987) Directed Ig class switch recombination in activated murine B cells. EMBO J 6:1663–1671

Wuerffel RA, Nathan AT, Kenter AL (1990) Detection of an immunoglobulin switch region-specific DNA-binding protein in mitogen-stimulated mouse splenic B cells. Mol Cell Biol 10:1714–1718

Somatic Hypermutability

Matthias Wabl[1] and Charles Steinberg[2]

1 The Phenomenon

In his book "*What is life?*," which was so influential at the time, Erwin Schrödinger (1944) mused, "In order to be suitable material for the work of natural selection, mutations must be rare events." That is, stability is one of the most important properties of the gene. A decade and a half later, the visionary Joshua Lederberg (1959) was musing about the origin of antibody diversity. He proposed that the instability of genes, i.e., from somatic mutation, was crucial in this process.

The duality of genetic stability and gene mutability is a recurrent theme in biology. On the one hand, genetic stability is the limiting factor in how big a genome can be (Eigen 1993). On the other hand, without mutability, there would be no organic evolution, and life as we know it today would not exist. This duality is apparent during the development of a single organism as well as in evolutionary time. On the one hand, without a high degree of genomic stability, large multicellular organisms could not exist, and such stability may be the limiting factor in the lifespans of vertebrates (Wabl et al. 1994). On the other hand, without mutability, the immune system of vertebrates would not function.

---

[1] Department of Microbiology and Immunology, University of California, San Francisco, CA 94143-0670, USA
[2] Basel Institute for Immunology, Postfach, 4005 Basel, Switzerland

Lest anyone protests that the last sentence of the preceding paragraph is hyperbole, we would like to point out that V(D)J rearrangement (reviewed in LIEBER 1992) and the heavy chain (H) class switch (reviewed in HARRIMAN et al. 1993) both involve the deletion of DNA from the chromosome. Deletions are mutations by any standards, and since there are at least three of them in every lymphocyte, T as well as B, one might even call it hypermutation. Junctional diversity arises from imprecision in the V(D)J joing process that gives rise to addition/deletion mutants at a frequency also deserving the epithet "hypermutation". And gene conversion, which is the principal basis of diversification of immunoglobulins in birds, and perhaps in some mammals, is also a form of hypermutability. But in the remainder of this paper, we shall confine ourselves to the phenomenon of hypermutation in the sense that it has become established in the jargon of immunologists – somatic point mutations that arise at a high rate in the segments encoding the variable (V) regions of immunoglobulins.

From protein sequences of mouse λ light chains produced by myelomas, Weigert and Cohn (WEIGERT et al. 1970) deduced that somatic point mutations must be a cause of variability. This point was given a molecular biological foundation by Tonegawa and colleagues (BERNARD et al. 1978). However, it was argued by some that these mutations might have arisen during the extended period of time that the myelomas had been propagated outside of the host in which they first arose. In a stroke of genius, CUNNINGHAM and FORDHAM (1974) not only showed that somatic mutation contributes to antibody specificity, they also determined a rate of 10^{-3} per bp per cell generation, a rate of hypermutation that would be acceptable today. In his experiments, Cunningham micromanipulated plasmablasts with specificity for sheep red blood cells; they were identified by the plaques they caused in a lawn of sheep red blood cells in the presence of complement. When daughter cells of these plasmablasts were transferred onto a lawn of a mixture of red cells from two individual sheep, various plaque morphologies were observed: some lysed red cells of one sheep only, yielding turbid plaques, some lysed red cells from both sheep yielding clear plaques, and some reacted to one sheep and cross-reacted with the other sheep, yielding so-called sombrero plaques. The authors' interpretation was that the specificity had changed due to hypermutation. Critics argued that differences in amounts secreted over time or isotype switching might have given rise to the three plaque morphologies. Despite the caveats, this was the first determination of a mutation rate in an antibody gene. Seven years later, a spate of publications reported somatic mutations in hybridomas detectable by base sequencing (GEARHART et al. 1981; CREWS et al. 1981; SELSING and STORB 1981; GERSHENFELD et al. 1981; KIM et al. 1981; BOTHWELL et al. 1981). By 1981, the term "hypermutation" had become politically correct.

After immunization, and especially after repeated immunizations, the affinity of the antibodies in the serum increases dramatically with time – three or four orders of magnitude improvements are possible. This increase in the quality of serum antibody is called affinity maturation. Affinity maturation in the early part of the response has been attributed to the selection of un-stimulated clones of the primary repertoire, the repertoire generated by V(D)J rearrangement. As antigen becomes limiting, clones with higher affinity have a selective advantage. But the primary repertoire alone will rarely have the high affinities associated with hyperimmunization. Mutants with even higher affinities will be progressively generated by somatic hypermutation, and these will be selected pari passu. The prevailing view today is that the biological role of somatic hypermutation is solely to contribute to affinity maturation. And indeed, cells undergoing hypermutation are found in the germinal centers, where the B-cell response, in large part, takes place (JACOB et al. 1991; KELSOE and ZHENG 1993).

Today, hypermutation in the immunoglobulin genes is the paradigm of a site-specific, stage-specific, and lineage-specific mutator that generates point mutations. Let us briefly summarize the conventional view of events in hyper-mutation. Immunoglobulin genes are assembled in the bone marrow from gene segments – the heavy chain genes at the pro-B, and the light chain genes at the pre-B cell stage. When a functional immunoglobulin molecule is formed and inserted into the membrane, the pre-B cell becomes, by de-finition, a B cell, which travels through the peripheral blood to the spleen, lymph nodes, and other sites in the periphery. When the immature B cell meets a self-antigen, it is taken out of the functional pool, either by clonal deletion (i.e., physical destruction) (RUSSELL et al. 1991; CHEN et al. 1995; HARTLEY et al. 1991; SHOKAT and GOODNOW 1995) or anergy (i.e., functional shutdown) (GOODNOW et al. 1989; GOODNOW 1992), unless the antigen receptor is "edited" by replacing the light chain and/or heavy chain genes (NEMAZEE 1993; TIEGS et al. 1993; GAY et al. 1993; RADIC et al. 1993). Mature B cells police the periphery, and some of them come in contact with antigen. The antigen-specific cells home into the germinal centers of lymphoid follicles, where they proliferate, differentiate, and hypermutate. The architecture of the germinal centers pro-vides an intricate meeting place for the several cells involved in the immune response, and this seems to provide the most effective site for selecting mutant antibodies with higher affinities. The cells that accumulate mutations and acquire higher affinity to antigen are destined to become memory cells. During the primary response, no cells of clones from this pool will differentiate into plasma cells (LINTON et al. 1989; SPRENT 1994). Upon secondary stimulation by antigen, the memory cells proliferate and develop into plasma cells, which now secrete antibodies of higher affinity. Neither memory cells (SIEKEVITZ et al. 1987; KUPPERS et al. 1993) nor plasma cells accumulate any more mutations. In plasma cells, the immunoglobulin mutator apparently is shut off (MILSTEIN et al. 1978; WABL et al. 1985; BACHL and WABL 1996; VESCIO et al. 1995). This may also be the case in memory cells, but the lower likelihood of improved affinity by mutation yields few new mutations to be fixed.

In the H chain genes, the segments encoding the V region are the "epicenter" (GEARHART and BOGENHAGEN 1983; GEARHART 1993) of mutation, with the frequency decreasing progessively in both 5' and 3' directions. The area of hypermutation of about 2 kb includes the flanking regions. The 5' boundary near the promoter region is distinct; the 3' boundary near the enhancer region is more loosely defined (reviewed in GEARHART 1993). While the loci encoding H chains and κ chains behave similarly, at the locus encoding λ chains a segment of the major intron is as hypermutable as the complementarity-determining regions (GONZALEZ-FERNANDEZ et al. 1994). Within the hypermutable sequences, there are sites of even higher intrinsic mutability than their neighbors, so-called hotspots (BEREK et al. 1985; BEREK and MILSTEIN 1987; LEVY et al. 1988; BETZ et al. 1993). Consensus sequence motifs can be found for the hotspots (ROGOZIN and KOLCHANOW 1992; BETZ et al. 1993; YELAMOS et al. 1995).

While the highest frequency of mutations is found in and around the segments encoding V regions, sequences that allow high-level expression of immunoglobulin genes are not enough for hypermutability. For the κ (SHARPE et al. 1991; BETZ et al. 1994) and H (BACHL and WABL 1996) chain loci, both the major intron enhancer and the remote 3' enhancer are necessary; λ chain genes also need other as yet undefined elements (HENGSTSCHLAGER et al. 1994). Transcription or at least a promoter seem to be necessary, but it need not be the immunoglobulin promoter (BETZ et al. 1994). Surprisingly, the actual V exon is not necessary for hypermutation at the κ chain locus (YELAMOS et al. 1995).

2 The Experimental Systems

In the 1980s, studies on hypermutation in mice more or less followed a simple scheme. V regions from myelomas were sequenced and compared with their germline counterparts, or hybridomas were generated from immunized regular mice or immunized mice with immunoglobulin transgenes, as pioneered by the Storb laboratory for the light chain (RITCHIE et al. 1984) and Grosschedl and Baltimore for the heavy chain (GROSSCHEDL and BALTIMORE 1985). The V regions of these hybridomas were sequenced and compared with the germ line V, D, and J segments and with the flanking sequences, as well as with each other. In these studies, genealogical trees could be constructed, and a wealth of information about the distribution and the nature of mutations accumulated (SABLITZKY et al. 1985; RUDIKOFF et al. 1984; MANSER et al. 1984; MCKEAN et al. 1984; GRIFFITHS et al.1984; CLARKE et al. 1985; BEREK et al. 1985; SIEKEVITZ et al. 1987). Storb used animals where the role of DNA replication and transcription could be assessed. Neuberger's and Milstein's laboratory developed the concept of "passenger genes" – genes that are not translated and hence do not accumulate mutations selected by antigen (SHARPE et al. 1991). Cells that strongly bound peanut agglutinin (PNA) have accumulated many more mutations than cells that bind PNA weakly, and this fact has been exploited to study a large number of mutations (GONZALEZ-FERNANDEZ et al. 1993; YELAMOS et al. 1995).

Today, the dramatic progress in cloning techniques has allowed some shortcuts in the above scheme. That is, although the generation of hybridomas is still useful for many purposes, it is no longer absolutely essential. This has allowed hypermutationists to branch out into other animal species. Thus frog (WILSON et al.1992, 1995), shark (DU PASQUIER 1993; HINDS-FREY et al. 1993), sheep (REYNAUD et al. 1991, 1995), and rabbits (KNIGHT and BECKER 1990; KNIGHT et al. 1991; ALLEGRUCCI et al. 1991; WEINSTEIN et al. 1994) have all been studied to some degree and have contributed to our understanding of the phenomenon of hypermutation.

Nor have studies been confined to the conventional approach alone. Studies in situ, where cells picked from germinal centers are analyzed, confirmed that hypermutation is active during cell proliferation after antigenic stimulation (JACOB et al. 1991, 1993; JACOB and KELSOE 1992). V regions of cells from the other site in the spleen where B cells proliferate and differentiate, the periarteriolar lymphoid sheath-associated foci, were found to be mostly unmutated. Cells picked from histological sections also allowed Hodgkin and Reed-Sternberg cells to be identified as B cells and to be staged (KUPPERS et al. 1994).

Compared with the studies in the mouse, the studies in human beings have been less broad and less controllable by the investigator. However, follicular lymphomas have been studied extensively in vivo and in vitro (LEVY et al. 1988; BERINSTEIN et al. 1990). Cells from follicular lymphomas resemble the centrocytes of the secondary immune response – the cells where most somatic mutations accumulate because of a high rate of mutation and a strong selection for better binding. Cells from such tumors are all clonally related; they contain the same V_H rearrangement, but may express a light chain gene different from the original one. The cells continue to mutate their idiotopes at a high rate in vitro (BERINSTEIN et al. 1990). They can be and probably are being used for studies in which target sequences are assayed for mutability.

A system to delineate *cis* sequences that either target the hypermutation enzyme system or are necessary for it in other ways, e.g., for transcription, has been worked out in Gearhart's laboratory. It is an elegant system even though it does not yet replicate hypermutation in endogenous, unaltered immunoglobulin genes (UMAR et al.1991). Mice with preformed immunoglobulin transgenes with deletions that span putative target sequences and that may preclude transcription or translation are analyzed for mutants in a reporter gene. The reporter gene is a bacterial suppressor tRNA inserted into the major intron, and mutation frequencies can be measured in bacteria after transgene rescue. So far, large and small deletions were the only mutations detected in these transgenic mice; perhaps the addition of the 3' enhancer to the transgene probe will allow full mutator activity.

Even though the final goal is description of hypermutation in the whole animal, the study of the mutator components in transgenic animals is cumbersome and often involves a heroic effort. Thus, mutations at the immuno-globulin loci have also been studied in lymphocyte lines. On the one hand, as mentioned above, immunoglobulin genes of myelomas were compared with their germline counterpart gene segments. On the other hand, the actual process has been also studied in lines. Using, e.g., the compartmentalization test (VON BORSTEL et al. 1971), it is possible to accurately determine mutation rates (COFFINO and SCHARFF 1971; BAUMAL et al. 1973; WABL et al. 1985, 1987; MEYER et al.1986; SCHARFF et al. 1987; JÄCK and WABL 1987). Moreover, the *cis* sequences that are necessary can be determined in cell lines, as can the differentiation stages that support hypermutation (ZHU et al. 1995; GREEN et al. 1995; BACHL and WABL 1996). This approach will become especially important in the search for *trans*-acting factors. Early studies with endogenous genes in lines were recently extended by transgenic approaches. The laboratory of Scharff and our own laboratory described μ gene constructs with all the *cis*-acting elements necessary for hypermutation comparable to the rate of the endogenous gene segments encoding the V region. The activity of the mutator does not seem to depend strongly on the position of the transfected gene in the genome. However, there are some differences in the composition and behavior of the constructs. The construct of the Scharff laboratory described in the original publication contained no 3′ enhancer. Our construct contained the 3′ κ enhancer; when the enhancer was left out, reversion frequencies dropped by two orders of magnitude (J. BACHL and M. WABL, manuscript in preparation). Furthermore, our construct did not hypermutate in a plasma-cytoma, while their published results were obtained in a plasmacytoma and in lines derived by fusing a pre-B cell and a myeloma. Due to methodological differences, the results cannot be compared directly in quantitative terms, but the effect on our hypermutable construct was at least two orders of magnitude greater than on our control construct, while the difference between V and C in their case was about fivefold. When a 3′ enhancer was added to the Scharff construct, it behaved essentially like ours (M. SCHARFF, personal communication).

3 Some Questions

Thus far, we have attempted to present the history and the current state of hypermutation research. We have deliberately attempted to avoid controversial issues. In this section, we will discuss what we consider to be open questions. We make no claim for completeness in the list of questions we cover. Moreover, the detail in which these questions are discussed is very uneven; the reader will have little trouble ascertaining which questions the authors are most con-cerned with at the moment.

3.1 Is Affinity Maturation the Only Raison d'Etre for Hypermutation?

Receptor Editing. We should keep an open mind for the possibility that hypermutation also contributes to the virgin B cell repertoire. "Editing" of B cell receptors to self-antigens, for example by L chain or by V_H replacement, is becoming increasingly accepted (see above). There is no a priori reason why such editing should not also result from hypermutation. In fact, this is precisely how a latterday immunologist would describe the now classical theory of the somatic generation of immune diversity of Niels Jerne (JERNE 1971). Current experimental evidence points to a low frequency of mutants in the primary repertoire (GU et al. 1991). However, it is unfair to compare the frequency in receptors selected for better antigen binding with the frequency in receptors selected against binding to (self-)antigen. So far, most studies involve antibody genes that have been selected for encoding a given specificity. Surely there is a much greater chance of abrogating binding by a single point mutation than of achieving better binding. Indeed, this was borne out by experiments in which mutant antiphosphorylcholine antibodies were generated by random mutagenesis in vitro. In none of the 46 mutants tested was the affinity increased, but in one-tenth of the antibodies with one amino-acid substitution, and in over half of the ones with two, the affinity was decreased (CHEN et al. 1992). Moreover, there is good evidence that hypermutation plays a role in increasing the naive immune repertoire in the sheep, even in the absence of antigen altogether (REYNAUD et al. 1995).

Memory. A side issue, not directly relevant to the process of affinity maturation is the question of how memory cells are put aside. The intuitive way to explain B cell memory would be that a B cell clone with antigen affinity above a certain threshold is triggered, expands, and differentiates into both plasma cells and memory cells; i.e., antibodies of the late primary response would be the fast line of defense in a second challenge of the same antigen. However, the current view is just the opposite. It is thought that there are two kinds of B cells. Upon antigenic stimulation, one sort differentiates into plasma cells, but does not give rise to memory cells, while the other sort differentiates into memory cells, but does not give rise to plasma cells. Of course, hypermutation would benefit immunological memory in both models. However, in the two-lineage model, hypermutation would not necessarily favor the survivors of a first microbial attack, since the fine-tuning of the response has not been tested for its biological consequence. It is as if we were to meet someone, but would better remember his brother, whom we had not met.

3.2 Is the T Cell Antigen Receptor Hypermutable?

A controversial isssue is the question of hypermutation in the T cell antigen receptor genes. Hypermutation has been both postulated and rejected on theoretical grounds. But in hundreds of sequences, there is no evidence for somatic point mutants. The counterargument is that people have not looked in the right places. There is one report of hypermutation in T cells in germinal centers (ZHENG et al. 1994). As these results have been challenged on experimental grounds (BACHL and WABL 1995), the jury is still out on this issue.

3.3 Is Hypermutation Really Hyperconversion?

Published work on the generation of antibody diversity in the chicken was so beautifully executed that there was no doubt about hyperconversion being the cause of diversity in both H and L chains (REYNAUD et al. 1987, 1989; THOMPSON and NEIMAN 1987). Later it was shown that hyperconversion also occurs in rabbits (KNIGHT and BECKER 1990; KNIGHT et al. 1991; ALLEGRUCCI et al. 1991; WEINSTEIN et al. 1994); this elegantly solved the long-standing problem posed by variable region allotypes which has hounded immunogeneticists for so long. But it has also been proposed that hyperconversion is responsible for the high frequency of mutants at the immunoglobulin loci in mice and human beings. (MAIZELS 1989).

   This suggestion has met strong resistance, and there are two main arguements against it: (i) the apparent independence of point mutations, and (ii) the absence of donor sequences. There are only three V gene segments at the mouse the λ locus, and two of them differ only slightly from each other. These differences do not explain the mutational patterns. There are also no sequences in the genome homologous to the major introns, yet many mutations are found there in all three immunoglobulin loci. Recent results in Milstein's and Neuberger's laboratories ought to put to rest speculations that all hypermutation is really hyperconversion. Bacterial sequences can replace the V exon as a target for hypermutation (YELAMOS et al. 1995). Of course, it is not excluded that hyperconversion also contributes to antibody diversity in mice and human and human beings. However, in the one case where experimental evidence has been reported, a putatively converted V exon in a transgene might have been an endogenous gene (XU and SELSING 1994).

3.4 Does Hypermutation Depend on Transcription?

As mentioned above, the 5' boundary for mutations is the promoter region, suggesting that transcription is necessary for hypermutation, or at least that factors of the transcriptional machinery are part of the mutator. In mammals as well as bacteria, there is a DNA repair system coupled with transcription

(HANAWALT 1994). If the polymerase happens upon a lesion in the strand being transcribed, it is corrected to match the untranscribed strand. If this mechanism were active in hypermutable B cells, and if it were applicable to mismatches of base pairs, this is just the sort of mechanism that would ensure the accumulation of mutations even in the absence of mitosis. The stalling of RNA polymerase II at a natural sequence-dependent pause site (hotspot) may also be sufficient to initiate a "gratuitous transcription-coupled repair" (HANAWALT 1994), and the repeated generation of repair patches at lesion-free sites might lead to higher levels of spontaneous mutagenesis.

A dependence of hypermutation on transcription would also be an attractive hypothesis to explain strand discrimination. At least it would be if the same strand were always preferentially affected. This, however, does not seem to be the case (YELAMOS et al. 1995). Nevertheless, hypermutation has been linked to transcription and DNA replication (ROGERSON et al. 1991).

3.5 What *cis*-Acting Sequences Are Needed?

The transgenic systems make a detailed analysis of the *cis* sequences needed for hypermutation possible. As described above, the large intron enhancer and the 3' enhancer are needed, but not a particular immunoglobulin promoter sequence or the V exon sequences. Further narrowing down the core sequences will be achieved by the same type of experiments that allowed defining the elements so far.

3.6 What *trans*-Acting Factors Are Needed?

The question of which *trans*-acting factors are needed can be easily summarized by saying that nothing is known for sure. Several intriguing hypotheses to explain hypermutation have been put forward (e.g., MANSER 1990; ROGERSON et al. 1991; GEARHART 1993) since BRENNER and MILSTEIN (1966) first proposed that an enzyme complex would actively place mutations and thus cause hypermutation at the immunoglobulin locus. While the Brenner-Milstein model does not explain the facts of hypermutation as the word is used today, it bears an uncanny resemblance to the way in which untemplated nucleotides, so-called N regions, are added at the DJ and VD junctions during immunoglobulin heavy chain gene rearrangement (DESIDERIO et al. 1984), a process that was discovered almost 20 years after the model was proposed. The enzyme terminal deoxynucleotidyl transferase (TdT), which is responsible for the N regions, has been found to be expressed only in pre-B and pre-T cells, but germinal center B cells do not seem to have been checked.

In our view, one component of the mutator is an enzyme that recognizes specific sequences at the immunoglobulin loci and makes single-strand breaks at or near these sites, i.e., a nicking enzyme, similar to that postulated for repair of damaged DNA. The nicked DNA segment would be nibbled back, resynthesized, and repaired, only to be nicked again. Another component would be a "repressor" of some sort that would at least partially suspend proofreading during the period of active hypermutation. Thus, the repair process itself would introduce mutations at an elevated frequency during the resynthesis phase induced by the putative site-specific nicking enzyme. Hypermutation would continue until at least one of three events transpired: (i) the recognition sequences for the nicking enzymes themselves are mutated, (ii) the nicking enzyme is turned off, or (iii) proofreading is turned back on. This model relies on the general repair system, with just one additional enzyme, the sequence-specific nickase, and one "repressor" to abrogate efficient proofreading.

It ought to be obvious that attention to the rapidly expanding field of mammalian repair enzymes will benefit the field of immunoglobulin hypermutation. For example, mice with a null mutation in the DNA mismatch repair gene *PMS2* have been recently reported (BAKER et al. 1995). It will be interesting to see whether these mice show more (or maybe even fewer!) mutations at the immunoglobulin loci.

3.7 Are There Hypermutable Sequence Motifs?

Knowledge about the nucleotide sequences affected might give us a clue as to the biochemical mechanism of hypermutation. Different reactivities of bases, neighboring sequences, and other factors known from mutation research in bacteria and phages must have an influence. Indeed, hotspots of mutability have been recognized (BEREK et al. 1985; BEREK and MILSTEIN 1987; LEVY et al. 1988; BETZ et al. 1993), and common sequence motifs have been deduced (ROGOZIN and KOLCHANOW 1992; BETZ et al. 1993; YELAMOS et al. 1995). When the influence of antigen selection is not excluded in the analysis of mutations, then it seems that the sequences RGYW (where R is a purine, Y a pyrimidine, and W is either A or T) and TAA are conducive to hypermutation (ROGOZIN and KOLCHANOW 1992). However, given that there are only four different bases, these sequences are not all that uncommon. When the intrinsic mutability – i.e., with no influence of antigen selection – was assessed, the sequence motifs CAGCT, GAGCTT, and AAGTT were deduced (BETZ et al. 1993; YELAMOS et al. 1995). The hot bases in these motifs are all G or C. At the lambda locus of the mouse, hotspots seem to be contained in palindromes; no consensus sequence is immediately obvious, but all 11 hot base pairs reported are at a G-C base pair (GONZALEZ-FERNANDEZ et al. 1994). While adjacent sequences must have an effect on hypermutability, we feel that the emphasis should be placed on the bases that are hypermutable, a question that we will discuss at some length in Sect. 3.8 below.

3.8 Are G-C Base Pairs Preferentially Mutated?

G-C Base Pair Preference in Frogs. In the clawed frog *Xenopus*, more than 90% of all somatic point mutations at the immunoglobulin heavy chain locus affect G-C base pairs (WILSON et al. 1992). The authors of the *Xenopus* study postulate that the primary action of the mutator is biased towards G-C and against T-A in mice as well as in frogs, but that strong antigen selection for mutants with better affinity obscures the original spectrum of mutants in mice. Affinity maturation, which is driven by antigen selection, is poor in frogs, perhaps because they lack germinal centers, and, as a consequence, the high percentage of mutations at G-C base pairs is preserved. The same preference has been observed in *Xenopus* tadpoles (WILSON et al. 1995) and in the shark *Heterodontus* (DU PASQUIER 1993).

G-C Base Pair Preference of Unselected Mutations at Nonimmunoglobulin Loci in Mammals. The preference for G-C might be a general feature of unselected mutations in mammals as well as frogs and in genes unrelated to immunoglobulins. Using published spectra of spontaneous mutations in seven distinct experimental systems in human beings and Chinese hamsters (ZHANG et al. 1992), WILSON et al. (1992) analyzed the relationship of the G-C/A-T mutation ratio against the percentage of point mutations. They argued that the fraction of point mutations is a measure of of the stringency of selection in the seven systems, all of which involve loss of activity of the enzymes hypoxanthine-guanine and adenine phosphoribosyl transferases. As can be seen from Fig. 1

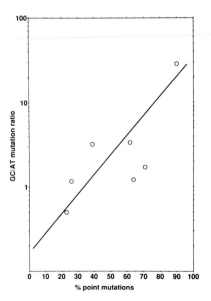

Fig. 1. Log GC/AT mutation ratio versus percent point mutations. The GC/AT mutation ratio is the ratio of cases in which the mutation arose from altering a GC base pair to cases in which the mutation arose from altering an AT base pair. This figure is taken from WILSON et al. (1993), each point corresponds to one column of data in Table 3 of ZHANG et al. (1992)

(taken from Fig. 7 of WILSON et al. 1992), the log G-C/A-T mutation ratio is positively correlated with the percent point mutations, hence negatively correlated with the stringency of selection. The correlation coefficient is 0.75, which means that somewhat more than half of the variability in log G-C/A-T can be accounted for by variability in the fraction of point mutations. G-C/A-T ratios of 20 or 30 are attained for 90%–100% point mutaions.

Nucleotide preferences in mouse and human immunoglobulin genes. Little overall bias for mutations of G-C base pairs is apparent in mouse hybridoma sequences (GEARHART 1993) and in the hypermutated nonimmunoglobulin sequences in transgenic mice (YELAMOS et al. 1995). However, it is interesting that within the motifs CAGCT and AAGTT (BETZ et al. 1993), GAGCTT (YELAMOS et al. 1995) and the small palindromes around the hotspots in the Vλ1 gene (GONZALEZ-FERNANDEZ et al. 1994) it is the G-C base pair that is the hotspot. One can argue that it is the motif, and not the G-C base pair that is the target here (see Sect. 3.7 above), but then one might also ask why the motif does not contain an A-T base pair in the center.

BETZ et al. (1993), classified mutational hotspots at the heavy and light chain loci of mice and humans into "selected" and "intrinsic," i.e., into those that seemed to have been selected by antigen and those that seemed to have not been selected by antigen. While there is no obvious preference for G-C base pairs among the selected hotspots, 12 of the 14 unselected hotspots, with 111 of the 123 mutations, are at G-C base pairs. This striking finding is in line with the interpretation that the primary specificity of the system is biased towards mutation of G-C base pairs, but this specificity is obscured by antigenic selection. However, it is also possible that the immunoglobulin mutator system in mammals consists of several mutators, of which the G-C mutator, as present in frogs, is only one.

Because antigenic selection is clearly not involved, analysis of the mutants found in the 18–81 cell line, might be expected to give more unequivocal results. Indeed, in each of the 19 mutations we have sequenced in this line, a G-C base pair was involved. As the mutations arose in separate culture wells that had been seeded with a single nonrevertant cell, we can be quite sure that they represent independent mutational events. Thus, the immunoglobulin mutator, at least in 18–81, affects G-C pairs. Of course, the fact that G-C base pairs are preferred does not necessarily mean that all G-C pairs are targeted with equal likelihood; nor does it exclude that other sequence requirements might be necessary. The TAG codon in these studies is not contained in the sequence CAGCT, but it is included in RGYW.

Scharff's laboratory reported an approach similar to ours in 18–81, without finding a bias for G-C base pairs (ZHU et al. 1995; GREEN et al. 1995). There were, however, significant differences in characteristics between the constructs and the cell lines used (see Sect. 2 above).

4 Conclusion

We can now look back at 25 years of experimental evidence for hypermutation in immunoglobulin genes. Although a rate of spontaneous mutation has never been measured at nonimmunoglobulin loci, the rate for hypermutation is thought to be a million times higher, varying somewhat over the V regions and its flanking sequences. Broad tendencies of the process have been mapped, and *cis*-sequences necessary for hypermutation have been defined. The challenge of today is to define how these *cis*-acting factors interact with *trans*-acting factors yet to be found.

Acknowledgments. We thank Michael Liskay and Juergen Bache for discussions on DNA repair and R.C. von Borstel for critical reading of the manuscript. This work was supported by NIH grant 1R01 GM37699 and by funds from the Markey Trust. The Basel Institute for Immunology was founded and is sponsored by F.-Hoffmann-La Roche, Basel, Switzerland.

References

Allegrucci M, Young-Cooper GO, Alexander CB, Newman BA, Mage RG (1991) Preferential rearrangement in normal rabbits of the 3′ VHa allotype gene that is deleted in Alicia mutants; somatic hypermutation/conversion may play a major role in generating the heterogeneity of rabbit heavy chain variable region sequences. Eur J Immunol 21:411–417

Bachl J, Wabl M (1995) Hypermutation in T cells questioned. Nature 375:285–286

Bachl J, Wabl M (1996) An immunoglobulin mutator that preferentially targets G-C base pairs. Proc Natl Acad Sci USA 93:851–855

Baker SM, Bronner CE, Zhang L, Plug AW, Robatzek M, Warren G, Elliott EA,Y J, Ashley T, Arnheim N, Flavell RA, Liskay RM (1995) Male mice defective in the DNA mismatch repair gene *PMS2* exhibit abnormal chromosome synapsis in meiosis. Cell 82:309–319

Baumal R, Birshtein BK, Coffino P, Scharff MD (1973) Mutations in immunoglobulin-producing mouse myeloma cells. Science 182:164–166

Berek C, Milstein C (1987) Mutation drift and repertoire shift in the maturation of the immune response. Immunol Rev 96:23–41

Berek C, Griffiths GM, Milstein C (1985) Molecular events during maturation of the immune response to oxazolone. Nature 316:412–418

Berinstein N, Campbell MJ, Lam K, Carswell C, Levy S, Levy R (1990) Idiotypic variation in a human B lymphoma cell line. J Immunol 144:752–758

Bernard O, Hozumi N, Tonegawa S (1978) Sequences of mouse immunoglobulin light chain genes before and after somatic changes. Cell 15:1133–1144

Betz AG, Rada C, Pannell R, Milstein C, Neuberger MS (1993) Passenger transgenes reveal intrinsic specificity of the antibody hypermutation mechanism: clustering, polarity, and specific hotspots. Proc Natl Acad Sci USA 90:2385–2388

Betz AG, Milstein C, Gonzalez-Fernandez A, Pannell R, Larson T, Neuberger MS (1994) Elements regulating somatic hypermutation of an immunoglobulin kappa gene: critical role for the intron enhancer/matrix attachment region. Cell 77:239–248

Bothwell ALM, Paskind M, Reth M, Imanishi-Kari T, Rajewsky K, Baltimore D (1981) Heavy chain variable region contribution to the NPb family of antibodies:somatic mutation evident in a gamma 2a variable region. Cell 24:625–637

Brenner S, Milstein C (1966) Origin of antibody variation. Nature 211:242–243

Chen C, Roberts VA, Rittenberg MB (1992) Generation and analysis of random point mutations in an antibody CDR2 sequence: many mutated antibodies lose their ability to bind antigen. J Exp Med 176:855–866

Chen C, Nagy Z, Radic MZ, Hardy RR, Huszar D, Camper SA, Weigert M (1995) The site and stage of anti-DNA B-cell deletion. Nature 373:252–255

Clarke SH, Huppi K, Ruezinsky D, Staudt L, Gerhard W, Weigert M (1985) Inter- and intraclonal diversity in the antibody response to influenza hemagglutinin. J Exp Med 161:687–704

Coffino P, Scharff MD (1971) Rate of somatic mutation in immunoglobulin production by mouse myeloma cells. Proc Natl Acad Sci USA 68:219–223

Crews S, Griffin J, Huang H, Calame K, Hood L (1981) A single VH gene segment encodes the immune response to phosphorylcholine:somatic mutation is correlated with the class of the antibody. Cell 25:59–704

Cunningham AJ, Fordham SA (1974) Antibody cell daughters can produce antibody of different specificities. Nature 250:669–671

Desiderio SV, Yancopoulos GD, Paskind M, Thomas E, Boss MA, Landau N, Alt FW, Baltimore D (1984) Insertion of N regions into heavy-chain genes is correlated with expression of terminal deoxytransferase in B cells. Nature 311:752–755

Du Pasquier L (1993) Phylogeny of B-cell development. Curr Biol 5:185–193

Eigen M (1993) Viral quasi-species. Sci Am 269:42–49

Gay D, Saunders T, Camper S, Weigert M (1993) Receptor editing: an approach by autoreactive B cells to escape tolerance. J Exp Med 177:999–1008

Gearhart PJ (1993) Somatic mutation and affinity maturation. In: Paul WE (ed) Fundamental immunology, 3rd edn. Raven, New York, pp 865–885

Gearhart PJ, Bogenhagen, DF (1983) Clusters of point mutations are found exclusively around rearranged antibody variable genes. Proc Natl Acad Sci USA 80:3439–3443

Gearhart PJ, Johnson ND, Douglas R, Hood L (1981) IgG antibodies to phosphorylcholine exhibit more diversity than their IgM counterparts. Nature 291:29–34

Gershenfeld HK, Tsukamato A, Weissman IL, Joho R (1981) Somatic diversification is required to generate Vκ genes of MOPC511 and MOPC167 myeloma proteins. Proc Natl Acad Sci USA 78:7674–7678

Gonzalez-Fernandez A, Gupta SK, Pannell R, Neuberger MS, Milstein C (1994) Somatic mutation of immunoglobulin lambda chains: a segment of the major intron hypermutates as much as the complementarity-determining regions. Proc Natl Acad Sci USA 91:12614–12618

Goodnow CC (1992) Transgenic mice and analysis of B-cell tolerance. Ann Rev Immunol 10:489–518

Goodnow CC, Crosbie J, Jorgensen H, Brink RA, Basten A (1989) Induction of self-tolerance in mature peripheral B lymphocytes. Nature 342:385–391

Green NS, Rabinowitz JL, Zhu M, Kobrin BJ, Scharff MD (1995) Immunoglobulin variable region hypermutation in hybrids derived from a pre-B- and a myeloma cell line. Proc Natl Acad Sci USA 92:6304–6308

Griffiths GM, Berek C, Kaartinen M, Milstein C (1984) Somatic mutation and the maturation of immune response to 2-phenyloxazolone. Nature 312:271–275

Grosschedl R, Baltimore D (1985) Cell-type specificity of immunoglobulin gene expression is regulated by at least three DNA sequence elements. Cell 41:885–897

Gu H, Tarlinton D, Muller W, Rajewsky K, Forster I (1991) Most peripheral B cells in mice are ligand selected. J Exp Med 173:1357–1371

Hanawalt PC (1994) Transcription-coupled repair and human disease. Science 266:1957–1958

Harriman W, Völk H, Defranoux N, Wabl M (1993) Immunoglobulin class switch recombination. Annu Rev Immunol 11:361–384

Hartley SB, Crosbie J, Brink R, Kantor AB, Basten A, Goodnow CC (1991) Elimination from peripheral lymphoid tissues of self-reactive B lymphocytes recognizing membrane-bound antigens. Nature 353:765–769

Hengstschlager M, Williams M, Maizels N (1994) A lambda 1 transgene under the control of a heavy chain promoter and enhancer does not undergo somatic hypermutation. Eur J Immunol 24:1649–1656

Hinds-Frey KR, Nishikata H, Litman RT, Litman GW (1993) Somatic variation precedes extensive diversification of germline sequences and combinatorial joining in the evolution of immunoglobulin heavy chain diversity. J Exp Med 178:815–824

Jäck HM, Wabl M (1987). High rates of deletions in the constant region segment of the immunoglobulin mu gene. Proc Natl Acad Sci USA 84:4934–4938

Jacob J, Kelsoe G (1992) In situ studies of the primary immune response to (4-hydroxy-3-nitrophenyl)acetyl. II. A common clonal origin for periarteriolar lymphoid sheath-associated foci and germinal centers. J Exp Med 176:679–687

Jacob J, Kelsoe G, Rajewsky K, Weiss U (1991) Intraclonal generation of antibody mutants in germinal centres. Nature 354:389–392

Jacob J, Przylepa J, Miller C, Kelsoe G (1993) In situ studies of the primary immune response to (4-hydroxy-3-nitrophenyl)acetyl. III. The kinetics of V region mutation and selection in germinal center B cells. J Exp Med 178:1293–307

Jerne NK (1971) The somatic generation of immune recognition. Eur J Immunol 1:1–9

Kelsoe G, Zheng B (1993) Sites of B-cell activation in vivo. Curr Opin Immunol 5:418–422

Kim S, Davis M, Sinn E, Patten P, Hood L (1981) Antibody diversity: somatic hypermutation of rearranged VH genes. Cell 27:573–581

Knight KL, Becker RS (1990) Molecular basis of the allelic inheritance of rabbit immunoglobulin VH allotypes:implications for the generation of antibody diversity. Cell 60:963–970

Knight KL, Becker RS, DiPietro LA (1991) Restricted utilization of germ-line VH genes in rabbits:implications for inheritance of VH allotypes and generation of antibody diversity. Adv Exp Med Biol 292:235–424

Kuppers R, Zhao M, Hansmann ML, Rajewsky K (1993) Tracing B cell development in human germinal centres by molecular analysis of single cells picked from histological sections. EMBO J 12:4955–4967

Kuppers R, Rajewsky K, Zhao M, Simons G, Laumann R, Fischer R, Hansmann ML (1994) Hodgkin disease: Hodgkin and Reed-Sternberg cells picked from histological sections show clonal immunoglobulin gene rearrangements and appear to be derived from B cells at various stages of development. Proc Natl Acad Sci USA 91:10962–10966

Lederberg J (1959). Genes and antibodies. Do antigens bear instructions for antibody specifity or do they select cell lines that arise by mutation? Science 129:1649–1653

Levy S, Mendel E, Kon S, Avnur Z, Levy R (1988) Mutational hot spots in Ig V region genes of human follicular lymphomas. J Exp Med 168:475–89

Lieber MR (1992) The mechanism of V(D)J recombination: a balance of diversity, specificity, and stability. Cell 70:873–876

Linton PJL, Decker DJ, Klinman NR (1989) Primary antibody-forming cells and secondary B cells are generated from separate precursor cell subpopulations. Cell 59:1049–1059

Maizels N (1989) Might gene conversion be the mechanism of somatic hypermutation of mammalian immunoglobulin genes? Trends Genet 5:4–8

Manser T (1990) The efficiency of antibody affinity maturation:can the rate of B-cell division be limiting? Immunol Today 11:305–308

Manser T, Huang SY, Gefter ML (1984) Influence of clonal selection on the expression of immunoglobulin variable region genes. Science 226:1283–1288

McKean D, Huppi K, Bell M, Staudt L, Gerhard W, Weigert M (1984) Generation of antibody diversity in the immune response of BALB/c mice to influenza virus hemagglutinin. Proc Natl Acad Sci USA 81:3180–3184

Meyer J, Jäck HM, Ellis N, Wabl M (1986) High rate of somatic point mutation in vitro in and near the variable-region segment of an immunoglobulin heavy chain gene. Proc Natl Acad Sci USA 83:6950–6953

Milstein C, Adetugbo K, Cowan NJ, Kohler G, Secher DS (1978) Expression of antibody genes in tissue culture:structural mutants and hybrid cells. National Cancer Institute Monographs 48:321–330

Nemazee D (1993) Promotion and prevention of autoimmunity by B lymphocytes. Curr Opin Immunol 5:866–72

Radic MZ, Erikson J, Litwin S, Weigert M (1993) B lymphocytes may escape tolerance by revising their antigen receptors. J Exp Med 177:1165–1173

Reynaud CA, Anquez V, Grimal H, Weill JC (1987) A hyperconversion mechanism generates the chicken light chain preimmune repertoire. Cell 48:379–88

Reynaud CA, Dahan A, Anquez V, Weill JC (1989) Somatic hyperconversion diversifies the single Vh gene of the chicken with a high incidence in the D region. Cell, 59:171–183

Reynaud CA, Mackay CR, Muller RG, Weill JC (1991) Somatic generation of diversity in a mammalian primary lymphoid organ:the sheep ileal Peyer's patches. Cell 64:995–1005

Reynaud CA, Garcia C, Hein WR, Weill JC (1995) Hypermutation generating the sheep immunoglobulin repertoire is an antigen-independent process. Cell 80:115–125

Ritchie KA, Brinster RL, Storb U (1984) Allelic exclusion and control of endogenous immunoglobulin gene rearrangement in kappa transgenic mice. Nature 312:517–520

Rogerson B, Hackett J Jr, Peters A, Haasch D, Storb U (1991) Mutation pattern of immunoglobulin transgenes is compatible with a model of somatic hypermutation in which targeting of the mutator is linked to the direction of DNA replication. EMBO J 10:4331–4341

Rogozin IB, Kolchanov NA (1992) Somatic hypermutagenesis in immunoglobulin genes. II. Influence of neighbouring base sequences on mutagenesis. Biochim Biophys Acta 1171:11–18

Rudikoff S, Pawlita M, Pumphrey J, Heller M (1984) Somatic diversification of immunoglobulins. Proc Natl Acad Sci USA 81:2162–2165

Russell DM, Dembic Z, Morahan G, Miller JF, Burki K, Nemazee D (1991) Peripheral deletion of self-reactive B cells. Nature 354:308–311

Sablitzky F, Wildner G, Rajewsky K (1985) Somatic mutation and clonal expansion of B cells in an antigen-driven immune response. EMBO J. 4:345–350

Scharff MD, Aguila HL, Behar SM, Chien NC, DePinho R, French DL, Pollock RR, Shin SU (1987) Studies on the somatic instability of immunoglobulin genes in vivo and in cultured cells. Immunol Rev 96:75–90

Schrödinger E (1944) What is life? Cambridge University Press, Cambridge

Selsing E, Storb U (1981) Somatic mutation of immunoglobulin light-chain variable-region genes. Cell 25:47–58

Sharpe MJ, Milstein C, Jarvis JM, Neuberger MS (1991) Somatic hypermutation of immunoglobulin κ L chain may depend on sequences 3' of C kappa and occurs on passenger transgenes. EMBO J 10:2139–2145

Shokat KM, Goodnow CC (1995) Antigen-induced B-cell death and elimination during germinal-centre immune responses. Nature 375:334–338

Siekevitz M, Kocks C, Rajewsky K, Dildrop R (1987) Analysis of somatic mutation and class switching in naive and memory B cells generating adoptive primary and secondary responses. Cell 48:757–770

Sprent J (1994) T and B memory cells. Cell 76:315–322

Tiegs SL, Russell DM, Nemazee D (1993) Receptor editing in self-reactive bone marrow B cells. J Exp Med 177:1009–1020

Thompson CB, Neiman PE (1987) Somatic diversification of the chicken immunoglobulin light chain gene is limited to the rearranged variable gene segment. Cell 48:369–378

Umar A, Schweitzer PA, Levy NS, Gearhart JD, Gearhart PJ (1991) Mutation in a reporter gene depends on proximity to and transcription of immunoglobulin variable transgenes. Proc Natl Acad Sci USA 88:4902–4906

Vescio RA, Cao J, Hong CH, Lee JC, Wu CH, Der Danielian M, Wu V, Newman R, Lichtenstein AK, Berenson, J (1995) Myeloma Ig heavy chain region sequences reveal prior antigenic selection and marked somatic mutation but no intraclonal diversity. J Immunol 155:2487–2497

von Borstel RC, Cain KT, Steinberg C (1971) Inheritance of spontaneous mutability in yeast. Genetics 69:17–27

Wabl M, Burrows PD, von Gabain A, Steinberg C (1985) Hypermutation at the immunoglobulin heavy chain locus in a pre-B-cell line. Proc Natl Acad Sci USA 82:479–482

Wabl M, Jäck HM, Meyer J, Beck-Engeser G, von Borstel R, Steinberg C (1987) Measurements of mutation rates in B lymphocytes. Immunol Rev 96:91–107

Wabl M, Harriman W, Fischer E, Steinberg C (1994) Can we outlive Methuselah? Cybernet Sys 25:373–387

Weigert MG, Cesari IM, Yonkovich SJ, Cohn M (1970) Variability in the lambda light chain sequences of mouse antibody. Nature 228:1045–1047

Weinstein PD, Anderson AO, Mage RG (1994) Rabbit IgH sequences in appendix germinal centers: VH diversification by gene conversion-like and hypermutation mechanisms. Immunity 1:647–659

Wilson M, Hsu E, Marcuz A, Courtet M, Du Pasquier L, Steinberg C (1992) What limits affinity maturation of antibodies in Xenopus – the rate of somatic mutation or the ability to select mutants? EMBO J 11:4337–4347

Wilson M, Marcuz A, Du Pasquier L (1995) Somatic mutations during an immune response in Xenopus tadpoles. Dev Immunol (in press)

Xu B, Selsing E (1994) Analysis of sequence transfers resembling gene conversion in a mouse antibody transgene. Science 265:1590–1593

Yelamos J, Klix N, Goyenechea B, Lozano F, Chui YL, Gonzalez-Fernandez A, Panell R, Neuberger MS, Milstein C (1995) Targeting of non-Ig sequences in place of the V segment by somatic hypermutation. Nature 376:225–229

Zhang LH, Vrieling H, van Zeeland AA, Jenssen D (1992) Spectrum of spontaneously occurring mutations in the hprt gene of V79 Chinese hamster cells. J Mol Biol 223:627–635

Zheng B, Xue W, Kelsoe G (1994) Locus-specific somatic hypermutation in germinal centre T cells. Nature 372:556–559

Zhu M, Rabinowitz JL, Green NS, Kobrin BJ, Scharff MD (1995) A well-differentiated B-cell line is permissive for somatic mutation of a transfected immunoglobulin heavy-chain gene. Proc Natl Acad Sci USA 92:2810–2814

Subject Index